Ship Performance

Ship Performance

SOME TECHNICAL AND COMMERCIAL ASPECTS

By

C. N. HUGHES, C.Eng., M.I.Mar.E.

|L|L|P|

LONDON   NEW YORK   HAMBURG   HONG KONG
LLOYD'S OF LONDON PRESS LTD.
1987

Lloyd's of London Press Ltd.
Sheepen Place, Colchester, Essex CO3 3LP
Great Britain

USA AND CANADA
Lloyd's of London Press Inc.
Suite 523, 611 Broadway
New York, NY 10012 USA

GERMANY
Lloyd's of London Press
PO Box 11 23 47, Deichstrasse 41
2000 Hamburg 11
West Germany

SOUTH-EAST ASIA
Lloyd's of London Press (Far East) Ltd.
903 Chung Nam Building
1 Lockhart Road, Wanchai
Hong Kong

©

C. N. Hughes

First published 1987

British Library Cataloguing in Publication Data

Hughes, C. N.
  Ship performance: some technical and
  commercial aspects.
  1. Ships——Performance
  I. Title
  623.8'2   VM149

ISBN 1−85044−149−9

Text typeset in 9pt on 11pt Century Textbook by
TND Serif Ltd., Hadleigh, Suffolk
Printed in Great Britain by
WBC Print Ltd, Bristol

Preface

The performance of a vessel is, of course, a very important aspect in the overall operating scenario affecting as it does both technical and commercial considerations. We will examine in this book some of the issues involved in the initial choice and subsequent operation of a vessel and also its main and auxiliary machinery.

Technical considerations have deliberately been pitched at a low level to generate a larger general interest in the subject-matter without recourse to higher mathematics. Commercial considerations are studied at some length because it is felt these have generally been neglected as evidenced by the lack of published matter on the subject.

Most of the material used in this work refers to what might loosely be called conventional vessels, i.e. bulk carriers and tankers, as these vessels probably account for over 90% of the world's tonnage.

July 1987   C. N. HUGHES

Table of contents

CHAPTER SEVEN   REMAINING COMPETITIVE BY OPTIMISING PERFORMANCE

CHAPTER EIGHT   PAST, PRESENT AND FUTURE

List of tables

List of figures

CHAPTER ONE

Technical considerations

1. Main propulsion system

The main propulsion system forms the major part of our study. It concerns not only the Main Engine but also the propeller whether it be fixed pitch, controllable pitch or ducted, to name but a few.

Generally the choice of Main Engine will be between slow speed and medium speed diesel. Both types have their supporters but the trend in recent years comes out in favour of slow speed diesels for our so-called conventional vessel. Assuming a slow speed diesel has been selected, we currently have two European and one Far Eastern design to choose from. All three designs have closely similar performances from a fuel consumption, in relation to horsepower, point of view. We will in future refer to this particular parameter as specific fuel consumption.

There are, however, various measures available to shipowners to enable them to improve performance of the propulsion system at the design stage. The most important of these options is probably to choose a derated engine having the highest possible stroke/bore ratio. It should, perhaps, be mentioned that derating is sometimes referred to as economy rating depending on the engine builder.

The principle behind derating concerns choosing a more powerful engine than that necessary for the selected ship's speed, and then operating it at a lower output but still using the maximum design combustion pressure. There is a twofold benefit operating in the derating mode in that the thermodynamic performance is improved by around 3% and propeller performance is similarly enhanced by operating at lower revolutions, this a well known principle in optimising propeller efficiency.

It should be mentioned that after choosing the derated option it is not possible to operate the engine at its full output without changing the propeller and fitting larger pumps, coolers, etc. One disadvantage of the derated engine is its high initial cost as the unused horsepower must, of course, be paid for by the shipowner, it being general practice that engines are sold on a horsepower basis.

This brings us to another important point which will be briefly mentioned here. Irrespective of whether a derated engine has been chosen, it is vitally important to determine what ship's speed is likely to be over the vessel's life-cycle. Ship's speed determines the output of the engine, and the advent of slow steaming in the early 1970s made most conventional vessels greatly overpowered. Many billions of dollars are locked up in unused horsepower, some of which will probably never be used.

A brief mention of the medium speed option will be made even though it is outside the scope of the subject-matter of this Chapter. Medium speed diesel engines generally operate better at their rather high design revolutions and, as far as is known, are not suitable candidates for the degree of derating possible on slow speed diesels. They are generally used in conjunction with gearboxes and clutches, and sometimes with controllable pitch propellers. Therefore a possibility of reducing propeller revolutions is instantly available, perhaps more so than with a derated long stroke slow speed engine, depending on the limitations of the propeller diameter, itself a function of many other considerations.

We can say that gearbox transmission losses could negate the possibility of any improvement due to reduced propeller revolutions leaving us with a straight thermodynamic efficiency contest which the slow speed diesel appears to be currently winning. Very many other considerations can be brought into the contest of slow speed versus medium speed diesels, but as we are only concerned with performance these will be left for others to discuss.

2. Underwater hull

The design of the underwater hull plays an important part in the overall performance of a vessel. There are many establishments around the world which tank test various designs submitted to them usually by a shipbuilder.

For bulk carriers and tankers a compromise is generally reached between maximising cargo cubic capacity without having too fine underwater lines. Fine lines are normally associated with high speed passenger ships and a different approach is necessary for these specialised vessels. Some very good designs have been produced in recent years. One of the most successful known to the author is the Burmister & Wain Panamax Bulker which has successfully been applied to their present series of product carriers.

So how is the performance of a vessel measured? There is an international test tank association which standardises the methods used in towing tank calculations, but this does not really help anybody but the most technically trained observer. In any case, the performance of a hull in a test tank and under seagoing conditions can be somewhat different, as many readers will know.

A well known method of measuring overall vessel performance in service is the non-dimensional fuel coefficient which is suitable for comparing vessels having similar particulars. The simple formula used calculating the coefficient is:—

$$\frac{D^{2/3} \, V^3}{C}$$

where D = displacement tonnage, V = ship's speed, C = fuel consumption (propulsion only or total consumption).

By introducing fuel consumption into the formula the machinery performance is also taken into consideration and a 3% less fuel consumption would increase the numerical value of the fuel coefficient by approximately the same amount.

It will also be seen that speed cubed is used to tie up with the well-known propeller law which uses this cube relationship as a basis for many calculations on hydrodynamic performance. This cube relationship illustrates the penal effect of speed on fuel consumption, and it can be shown that a 10% increase in speed requires a 30% increase in fuel consumption.

Some ship operators use the Admiralty Constant which uses horsepower instead of fuel consumption in the fuel coefficient formula. However, not many vessels are fitted with accurate meters for measuring horsepower, so a more practical approach using fuel consumption is preferred by the author. Strictly speaking the Admiralty Constant is a more accurate measurement of the ship's hull. If we substitute horsepower instead of fuel consumption the performance of the engine, whether it be a fuel-thirsty turbine or a modern diesel, is not taken into account. What sort of fuel coefficient or Admiralty Constant values can we expect in service? A lot depends on the type of vessel and expected figures for fully laden conditions using 9,600/K.cal/kg fuel are given in Table 1. For ballast conditions, the numerical values of both the Admiralty Constant and fuel coefficient are somewhat lower.

Table 1. Fuel coefficients and Admiralty Constants

VESSEL TYPE	FUEL COEFFICIENT		ADMIRALTY CONSTANT	
	Sea trials	*In service fairweather*	*Sea trials*	*In service fairweather*
VLCC/VLBC	190,000	170,000	600	540
Cape sized bulker	190,000	170,000	590	530
Panamax bulker/tanker	185,000	157,000	570	490
30,000 TDW bulker/tanker	150,000	135,000	395	350
16-year-old VLCC	137,000		535	

The vessels in the table are assumed to have the latest propulsion systems with specific fuel consumptions of around 127 gms/b.h.p./hr basis heavy fuel of net calorific value 9,600 K.cals/kg or 40.20 megajoules/kg, to use this recently introduced unit. Corrections can be made for older vessels by making allowance for changes in specific fuel consumption and propeller revolutions.

By way of example, we could take a 1970-built diesel VLCC which had a sea trial fuel coefficient of around 137,000, an Admiralty Constant of around 535 and a specific fuel consumption of 166.5 gms/b.h.p./hr at 9,600 K.cals with Main Engine revolutions of 100.

The increase in fuel coefficient over the modern vessel shown in the table is nearly 40%, but the improvement in the Admiralty Constant is only around 12%. Most of the 40% improvement in fuel coefficient is explained by a 31% improvement in specific fuel consumption and the remainder by the reduced RPM of the propeller coupled with improved aft-end geometry. The increase in Admiralty Constant is due to reduced RPM and aft-end geometry only. As a rough rule of thumb it is generally accepted that a 10% reduction in propeller revolutions equates to a 1% increase in speed, which is equivalent to a 3% reduction in power.

Using this approach, we can compare the 100 RPM of the older vessel and the 70 RPM of the modern vessel which gives a 43% reduction in RPM equal to a 12.9% reduction in power, closely that of the improvement in Admiralty Constant.

Differences between sea trials and in-service results are occasioned firstly by the so-called Sea Margin normally added to sea trials results to take account of weather and secondly, by the expected deterioration in hull and propeller roughness after entering service. Sea trials on bulkers are invariably carried out in the ballast condition and corrections are made for loaded displacement conditions, whereas tankers generally carry out their sea trials in loaded displacement condition and no displacement corrections are necessary.

Contractual conditions for carrying out sea trials usually specify calm weather and additionally various corrections are made for wind speed, water depth, tide, water density and temperature. All these add up to, perhaps, 15–20% increased power being necessary to maintain sea trials speed in service. A lot depends on the expected trading pattern of the vessel with its effect regarding sea state and weather conditions also hull fouling organisms. Only the shipowner can determine what figure to use, based on his experience.

Modern underwater paint systems have been especially formulated to cope with the current practice of extending dry dock intervals to combat high operating costs in a critical expense area. The use of the fuel coefficient and Admiralty Constant can detect a trend in performance fall-off and help shipowners decide if a dry docking, or underwater scrubbing operation, is necessary.

They are also useful for determining if a shipyard's design is a good performer or not. The main drawback to their unqualified use is that the cube relationship between speed and horsepower has, in some instances, been measured as high as the fifth power which can make direct comparisons a little dangerous.

3. "Bolt-on" improvers

With the advent of the progressive fuel cost hikes experienced starting from the Autumn of 1973 until the Spring of 1986, many innovative ideas to improve the performance of vessels have been introduced.

A. Slow steaming

Not strictly in this category is the widely practised slow steaming mode of operation. However, when carried out properly, slow steaming requires special fuel nozzles to be fitted which basically increase fuel injection pressures to sustain acceptable combustion conditions below that originally thought necessary in the original design.

Various operational measures of a protective nature were also introduced simply to counter the possible deleterious effect of continuous operation at ultra-low powers which have been allowed to drop to around 25% in suitable cases. In this category are included diesel VLCCs returning to the Arabian Gulf for an indefinite stay when worldscale levels of under 20 prevailed.

Other measures to obtain more dramatic improvements in slow steaming performance involve fitting new propellers and in the case of steam turbine vessels altering the gear ratio or turbine layout. This course of action enables much lower propeller revolutions to be used more economically, but still maintaining the turbine efficiency at almost full power output.

Modern diesel engines have fairly constant specific fuel consumptions throughout their operating range; steam turbines decidedly do not have this attractive feature, hence the need for major surgery to enhance the steam turbine's relatively poor performance — especially at low power output.

To improve the specific fuel consumption of older diesel engines at lower outputs or if slow steaming, the economy fuel pump was developed. This loosely follows the derating principle mentioned earlier and the economy fuel pump increases maximum combustion pressure at lower powers with a corresponding increase in thermodynamic performance of around 2–3% depending on the chosen power. The economy fuel pump loses its effect above 85% and below 40% output.

The practice of slow steaming, with or without the bolt-on benefits, is arguably the most cost-effective method of improving performance as measured by a miles per tonne of fuel basis.

B. Power take-offs

The generation of auxiliary power, normally in the form of electrical energy, has scope for improvement in the overall performance of the machinery within the vessel. Normally electrical energy is provided by independent diesel generators using a distillate fuel originally diesel oil, but more latterly blended or even heavy fuel.

A development gaining recent popularity is to use a power take-off from the Main Engine in the form of a shaft generator mounted directly on the intermediate shaft or free end of the crankshaft. Variations on this theme include hydraulic or geared drives taken from the crankshaft. Some versions have a novel revolution regulating system which permits less expensive voltage regulation control. This is an important consideration when a Main Engine of variable output is used to perform a duty normally requiring fairly constant revolutions.

The main benefit in using a power take-off is that the fuel used in the Main Engine, which — if of slow speed design — can tolerate fuel of quite low quality, is used to generate electrical energy whilst at sea. It should be remembered that the amount of electrical energy used must be deducted from the developed propulsion energy with a resultant loss of the ship's speed. Another benefit of using power take-offs is that maintenance costs of the diesel generator engines are reduced because they are only used in port or whilst manoeuvring.

The main disadvantage in power take-offs is the rather large capital cost, which probably can be justified when fuel costs are high but not when they are low. Each shipowner will have his own ideas on the cost-effectiveness of power take-offs.

It should perhaps be mentioned in closing that in a depressed shipbuilding market it is possible to obtain extras such as this at cost price. We must not forget that the universal trend to use lower cost crews makes achievement of the technical aspect necessary for highly technical innovations that much more difficult in service.

C. Exhaust gas recovery

As mentioned in the previous section, power take-offs do not make a positive contribution to performance improvement, in that the power used is redirected rather than conserved. This is not the case in exhaust gas recovery schemes which provide energy without detracting from the Main Engine output.

Most vessels have simple exhaust gas recovery schemes, generally in the form of economisers or boilers, which supply steam for all heating purposes by utilising the waste heat in the Main Engine exhaust gases. Such are the benefits of this scheme that in recent years it has been extended to recovering heat from the exhaust gases of diesel generators, albeit at a rather lower value than that from Main Engines.

In a sophisticated exhaust gas recovery scheme a much larger economiser is provided, and all the available heat in the gases is converted to steam. Sufficient steam is generated to drive a steam turbine coupled to an alternator which can supply all seagoing electrical demands. On vessels fitted with rather large powered Main Engines, alternators having outputs of around 1,000kW are not unknown.

Recent improvements in Main Engine thermal efficiencies, which now exceed 50%, have dealt a severe blow to these steam driven alternator schemes. This improved efficiency has manifested itself in lower Main Engine exhaust gas exit temperatures so the amount of heat capable of being recovered has dropped dramatically. The situation has been partially retrieved by the adoption of air-cooled turbochargers which have higher gas outlet temperatures and, therefore, have a higher heat input into the economiser allowing more steam to be generated.

Another innovation is the hot gas bypass which allows a certain amount of high temperature exhaust gas from the exhaust manifold to flow direct into the economiser. Unlike the air-cooled turbocharger, which has no effect on Main Engine performance, the hot gas bypass does represent a slight loss of Main Engine efficiency, albeit very small.

Exhaust gas turbochargers have also had their efficiency improved and any excess power available can be returned to the propeller shaft by introducing an additional power turbine, utilising a novel system of gears similar to that used in some of the power take-off schemes described earlier. The contribution from this turbocharger scheme is around 3-4% in suitably powered vessels.

It is even possible to have a combined system whereby the output from the power turbine can be directed to the Main Engine crankshaft or to an electrical alternator supplying the ship's electrical needs as shown in Figure 1.

Fig. 1. MAN-B & W power take-off with turbine

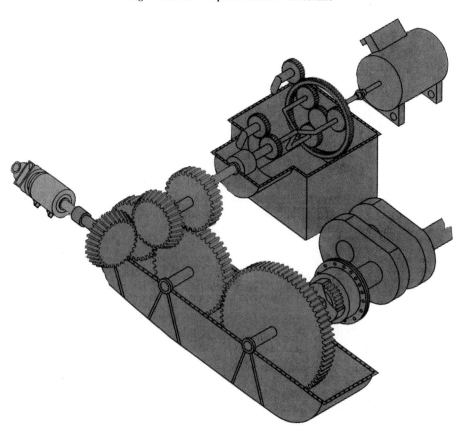

Sophisticated exhaust gas recovery systems are rather expensive but when fuel costs are pitched at what might be called their normal level an economic case can be put forward, providing a large enough Main Engine is available to generate sufficient waste heat from the exhaust gases or surplus power from the turbochargers. Retrofitting sophisticated exhaust gas systems is not normally an attractive proposition and their use is generally confined to new buildings.

D. Use of blended or residual fuel in diesel generators

As mentioned in a previous section, the generation of electrical power is normally by diesel generators using distillate fuel. For reasons known to themselves, the oil majors have always priced distillate fuel disproportionately high. The oil price hikes, starting in the 1970s, highlighted this anomaly and various designs for using a mixture of residual heavy fuel and distillate fuel appeared on the market.

Distillate fuel, normally diesel oil, has rather low viscosity and does not require heating. Residual heavy fuel, by having a high viscosity, requires substantial heating and when these fuels are mixed or blended a degree of heating is required. This is the main modification required when blended fuel is used and a fuel heating system, as well as a static or dynamic blender, must be provided to ensure the blend is adequately mixed and heated.

Because the resultant blend will contain some of the naturally occurring contaminants normally associated with residual heavy fuel, it may be necessary to upgrade some of the components in the diesel generator engines. These contaminants are not normally present in distillate fuel, and engines designed to run on this good quality fuel could run into operational problems — unless fitted with high grade exhaust valves and other modified components.

A logical progression is to use straight residual heavy fuel in the generator engines after experience gained running on blended fuel. Several manufacturers now market diesel generator engines for use on residual fuel and many "one fuel ships", as they are called, are in service.

Retrofitting existing plants to run on blended or residual fuel is sensitive to ruling fuel costs, and pay-back periods of around 12–36 months are often quoted. A disadvantage is the increased cost of maintenance and possibly increased lubricating oil costs which should be included in any calculations.

E. Upgrading existing Main Engines

The vast majority of Main Diesel Engines in service today have rather low thermal efficiencies when compared to the engines currently available. However, it does appear that we are reaching the practical limit, and it is doubtful if any significant increase in thermal efficiency above the 53% or so presently attainable will be made. This 53% should be compared with the 40% of the older engines and, although it would be possible to re-engine many vessels, this expensive exercise has rarely been undertaken in recent years. Compared with the 30% efficiency of the steam turbine even the older diesels look good, and many steam-to-diesel conversions have taken place. There exists a possibility of upgrading certain diesel engines and some of the options available are outlined.

One of these options is to convert from impulse turbocharging to constant pressure turbocharging which increases thermal efficiency by around 10%. It is quite a major operation to convert to constant pressure turbocharging — new turbochargers and a new exhaust manifold are necessary.

It is also possible to increase the stroke/bore ratio of some earlier engines by fitting cylinders of a smaller bore diameter. One disadvantage of this option is that the output of the engine is considerably reduced, which may not suit all shipowners' requirements.

Generally, both these operations are carried out in conjunction so the combined effect in improved performance almost approaches that of the latest engines.

F. Ducted and other propellers

The performance of conventional fixed pitch propellers has probably reached its limit and the only possibility left is to increase diameter with a corresponding reduction in revolutions and a corresponding increase in efficiency. There is a practical limit to the size to which a propeller diameter can go, mainly because of draught restrictions. Propeller efficiency can also be improved by using various means which are described below.

Flow straightening nozzle or wake improvement duct (Fig. 2)

This is fitted in front of the propeller and its purpose, as its name suggests, is to even out the flow of water into the propeller. It consists of a nozzle which is made in two halves and welded to the stern frame. Propeller efficiency can increase upwards of 6%, especially if high speed vessels are under consideration. The nozzles are not so effective when fitted to slow speed vessels.

Fig. 2.   Flow straightening nozzle or wake improvement duct.

Fig. 3.   Integrated duct

Integrated duct (Fig. 3)

This is mounted in front of the propeller and requires to be a quite massive structure compared to the flow straightening nozzle and is, therefore, rather expensive. The theory involved is to recapture some of the energy lost in turbulent forces and redirect it to the forward propulsive force. In suitable cases, gains of up to 8% increased efficiency can arise.

Reaction fins

These are fitted in front of the propeller and are sometimes fitted with a shroud to combat dynamic forces in this critical area. Their purpose is to improve the propulsive performance by inducing a swirl to the propeller inflow which will counter slipstream rotation, a source of efficiency loss. Gains of 4% to 8% are claimed by fitting these reaction fins. Retrofitting costs are rather high if a shroud is necessary but, if the reaction fins are designed for a new vessel, the shroud could probably be eliminated and the cost would, therefore, be much less.

Vane wheel (Fig. 4)

The vane wheel consists of what might be termed a free running propeller mounted directly behind the existing propeller. The vane wheel is larger in diameter than the main propeller and has a larger number of blades. The purpose of the vane wheel is to convert tangential forces in the propeller slipstream into additional thrust. Gains in propulsive efficiency of around 10% have been quoted and payback times of 12 months indicated. One problem is that not all existing single screw ships have aft-end arrangements capable of accommodating this rather large appendage without major restructuring.

Fig. 4.   Vane wheel

High skew

This type of propeller is becoming increasingly popular, but it has never been claimed to be a performance improver. It could be said that it permits the same blade area to be developed over a smaller diameter, which allows vessels with a draught restriction to be fitted with more effective propellers.

Nozzle rudder

The nozzle rudder is a unique combination of propeller duct and rudder. It has very good manoeuvrability, especially at low speed. No claims for any improvement in performance is claimed but it would appear to be more efficient than a normal propeller.

Tip vortex free

In its original form the tip vortex free propeller was used in conjunction with a duct and trial results indicated satisfactory savings — especially on VLCCs. More recently, it has been developed without a duct and renamed HEFA (High Efficiency Flow Adapted). The tips of the HEFA propeller are provided with end plates which permit a finite load to be developed in this part of the propeller, so having an advantage over a conventional propeller.

Asymmetrical stern

The aft-end lines of the vessel are offset to improve the water flow into the propeller and compensate for the side thrust generated by the direction of turning. It is not strictly a bolt-on performance improver, as it would be difficult and cost prohibitive to modify an existing vessel. Improvements in performance of around 8% are claimed and, if the asymmetrical stern is negotiated during contractual discussions, only negligible cost would be involved.

Controllable pitch

Controllable pitch propellers are generally used in cross-channel ferries and other vessels having a high manoeuvrability requirement coupled with multi-engine configurations. They have only rarely been used in our so-called conventional vessels which, of course, do not have such high manoeuvrability requirements.

Because of the large difference in the loaded and ballast draughts of conventional vessels, controllable pitch propellers do have a part to play as they can be adjusted to optimise the displacement condition. They are also useful when used in conjunction with a shaft generator as revolution control is made much easier. It would be difficult to calculate what improvement in performance would arise, as controllable pitch propellers are not specifically marketed for this reason. They are rather costly and are not particularly efficient, mainly due to their increased weight and reduced blade area, compared with a fixed pitch propeller.

Faced with this array of bolt-on performance improvers it is difficult for any shipowner to make a choice — especially as the drop in fuel costs, starting spring 1986, made most alternatives financially unattractive. As will be discussed later on in the book, a lot depends on whether the vessel is time chartered or not. However, as a rough guide to measure the financial attractiveness of some of the performance improvers mentioned, Table 2 will give some idea in order of magnitude for a typical Panamax bulker.

Table 2. **Performance improvers — sensitivity**

Performance improver	Capital cost	Percentage improvement		Recovery period in months with various fuel cost and steaming conditions							
	$	Slow steam	Full steam	$50 Slow	Full	$100 Slow	Full	$150 Slow	Full	$200 Slow	Full
Increased stroke/ bore — constant pressure	800,000	15	17	222	75	111	38	74	25	56	19
Constant pressure only	400,000	10	12	167	53	83	27	56	18	42	13
Vane wheel	200,000	6	10	139	32	69	16	46	11	35	8
Flow straightening nozzles	50,000	4	8	52	10	26	5	17	3	13	2
Economy fuel pumps	20,000	2	1	42	32	21	16	14	11	10	8

Capital costs are given in good faith at 1986 levels, but US$ cross-rates and geographical location of the work done could cause these to be somewhat different. The power of the

Main Engine also plays an important part in the decision-making process. The vessel chosen in the example has a rather high powered engine, and lower powered vessels would give less savings. A word of caution about the percentage improvements shown. These, generally, cannot be guaranteed by the manufacturers and it is for the shipowner to decide if the claims made are achievable in service.

4. Energy conservation

This subject is closely allied to that of performance improvers mentioned in the previous paragraph. For ease of reference, the items included in this paragraph will be those not of a large capital outlay.

A. Minimum ballast — optimal trim

Our conventional vessel will probably spend 50% of sea time loaded, and the other 50% in the ballast condition. There is a lot of scope for improving performance in the ballast condition particularly on the larger vessels. If we take as an example a Panamax bulker, the difference between the light and heavy ballast condition is something in the order of 10,000 tonnes. This represents a speed increase of around 0.2 knots if the vessel were operated wholly in the light ballast condition on ballast voyages. On larger vessels, the speed difference can be greater on account of the greater reduction in ballast capacity.

Of course safety considerations must prevail, especially when severe weather is forecast. Most shipmasters frown upon filling the deep tank (cargo hold) at sea, but a close study of the problem has shown that on many vessels it is possible to achieve a light ballast condition with the deep tank full — thus allaying fears of having to fill this tank in inclement weather. In the case of tankers MARPOL regulations relating to minimum draught must be considered and in the case of large bulkers registered in the UK, similar DOT recommendations should be observed.

Most vessels have the possibility of choosing an optimum trim by adjustment of loading or ballasting operations within certain limits. The only problem is knowing what this optimum trim actually is. Although there are various microprocessors on the market which allegedly calculate the answer for you, the technical aspects are not sound. Each vessel type will have its own characteristics which could change in different sea states, and a series of extensive — and expensive — tank tests would appear necessary before making any firm recommendations. The type of bulbous bow fitted would appear to have an influence on both aspects considered in this section. Only detailed analysis of service results can given an indication of what improvements are likely.

B. Adaptive steering modules

There are also microprocessors available which allegedly improve steering capabilities and, therefore, set a course that minimises the distance between ports. The theory is sound but, in the author's experience, the service results did not live up to the claims made. This is not to say that all steering modules would fall into the same bracket, but the subject is very difficult and measuring the results perhaps requires an approach beyond present measurement techniques.

C. Self-polishing underwater paint

The use of self-polishing copolymer underwater paint has spread in recent years; its main purpose is to allow extended periods between drydockings. Frictional resistance probably accounts for 70% of the total ship resistance, the balance being made up of wave, eddy and air resistance. Conventional underwater paint was originally based on the almost universally adopted annual drydocking. As drydock intervals were extended as a cost-cutting exercise, paints having more aggressive anti-fouling properties were developed.

The high fuel cost era, starting in 1973, highlighted the need to reduce frictional resistance which eventually led to self-polishing copolymer paint being marketed which has the dual function of combating marine growth and reducing frictional resistance. This latter property is achieved by using the action of the vessel proceeding through the water to slowly abrade the paint surface making it progressively smoother. Obviously, paint film thicknesses must be that much greater when using self-polishing paint and this, coupled with a higher product cost, makes the operation that much more expensive than conventional paint systems.

A very important factor is the roughness of the hull which must be maintained as smooth as possible. Latest thinking is to take regular hull roughness readings at each drydocking to monitor the situation. Shot blasting is really the only answer starting from when the ship is built and then maintaining damaged areas by blasting as necessary at subsequent drydockings. A sophisticated sacrificial cathodic protection system is a good idea or, if trading in areas where underwater damage is expected, ice for example, an impressed current cathodic system will greatly help control corrosion.

D. Underwater air lubrication

Towing tests carried out at the Feltham Tank in 1979 gave encouraging results when using, first, a simple plate and, latterly, a model. However, when the system was fitted on a vessel no significant reduction in hull resistance could be measured. It would appear that the theoretical reasoning is sound but the practical difficulties in obtaining an air film under seagoing conditions are insurmountable.

E. Propeller polishing

Propeller surfaces are as much likely to be affected by fouling as is the underwater hull, although corrosion is not normally a threat because of the non-ferrous metals used in propeller manufacture. It is important that the propeller blade surfaces are maintained as smooth as possible and this is normally carried out by discing at the drydock.

Specialist firms for carrying out this operation afloat have recently been formed and they claim that, by doing the work underwater, a better finish is obtained — a claim which has some substance. No significant increase in performance has yet been measured after several underwater polishing operations have been carried out by the author's company.

It is difficult to say if the propeller polishings carried out at drydocking are effective, as the maintenance work carried out on the underwater hull would tend to overshadow that carried out on the propeller.

F. Underwater scrubbing

This operation has been in practice many years and gained popularity in the early 1970s when it looked likely that the availability of large drydocks would be unable to meet the rising number of vessels of ever-increasing size entering service. This situation did not last for very long, but extended drydock intervals became the norm for economic rather than logistic reasons.

Many stations for carrying out underwater scrubbing operations are now available worldwide and can quickly and efficiently do the work. Video tapes showing before and after conditions are normally available. Performance improvement is usually dramatic in the days after the operation, but it generally falls off after a period which varies for reasons which are at present unknown. There is a possibility that the scrubbing brushes remove the anti-fouling ingredient in the paint which leaves the remaining surface more vulnerable to attack by marine growth. A further possibility is that the brushes scatter the spores of the organisms, thus promoting further growth.

G. Reduction of electrical load

Electricity costs money, especially when diesel oil costs were as high as they were in the early 1980s. To leave a 100 watt light burning would then cost around $50 per year if conventional diesel generators using diesel oil were in use. This gives some scope for conserving energy aboard ship by reducing the electrical load. An approach which has met with some success is to introduce an energy audit, whereby every item consuming electricity is examined with a view to either stopping it completely, limiting its output or using an alternative means of performing its function.

A prime example is a typical sea-water cooling system which normally comprises a main circulating pump and possibly two auxiliary pumps. It was found possible to stop one of the auxiliary pumps as the main pump could adequately supply the auxiliary system, mainly because the Main Engine was being operated in a slow steaming mode. In other cases, a simple cross-connection between main and auxiliary systems made it possible to stop an auxiliary pump.

Another idea with possibilities is to reduce the electrical frequency by slowing down the generators — this in turn slows down all the electric motors and reduces the output of whatever is being driven by the motors. When slow steaming this is quite acceptable: reducing the frequency causes no problems as long as it is increased to normal when full steaming is restored. During the power supply problems of the 1950s the Central Electricity Generating Board reduced the frequency of the entire grid system to conserve energy during peak demand periods.

Apart from switching off many lights, the audit was able to find various examples whereby other electrical consumers could be switched off. Hold ventilation fans, even radars, came into this category. It was not uncommon to find that reductions of 30kW in electrical load were possible at an annual cost saving of around $15,000 without any adverse effect. A cautionary note regarding reduction of frequency must point out that unless this is accomplished with a corresponding reduction in voltage overheating of electrical motors could occur.

H. Increase in deadweight

It is possible to increase the deadweight and hence earning capacity of many vessels, particularly larger-sized older bulkers by adopting the B-60 freeboard. As a typical example, a 1974-built CAPE sized bulker — constructed under the then existing freeboard regulations — had a summer deadweight of 123,132 tons. By carrying out certain work to the classification society's specification, it was possible to increase the summer deadweight to 129,193 tons, a very useful contribution to the vessel's earning capacity particularly in times of reasonable freight rates. There is a penalty to pay in increased wetted area leading to higher frictional losses with a corresponding loss in speed or increase in fuel consumption. But this is only in the fully laden condition. We also have to consider the capital cost of the conversion but generally, it would appear the benefits outweigh these penalties.

I. Other action

This section could also be called good housekeeping, as the subject-matter is mainly just that. Many apparently similar vessels perform quite differently, and the reasons for this will be examined.

One of the most common faults concerns the number of diesel generators in use at sea. Some vessels use two machines as a misguided security operation whereas one machine could adequately cope with the anticipated additional load. Using two machines costs many thousands of dollars per year in additional fuel consumption.

Correct maintenance of Main Engine temperatures is very important if thermal efficiency is to be kept at the design figures. Again, some vessels — as a misguided security measure — keep certain critical temperatures below the recommended value with resultant loss of efficiency.

The same applies to combustion pressures, especially as fuel quality variations are now frequently met with. The net effect of not maintaining these critical temperatures and pressures represents a loss of performance, manifesting itself in increased fuel consumption.

It has been found that some vessels still use diesel oil whilst manoeuvring, even though the fuel system has been designed for operation on heavy oil when entering and leaving port.

Probably the most important aspect is maintenance of the underwater hull. We have learnt that frictional resistance accounts for 70% of the total propulsive resistance. If this resistance is allowed to increase by not keeping the hull surfaces in reasonably good condition, then fuel consumption will rise dramatically. Choice of paint system, regularity of dry docking, geographical area of operation and weather conditions when applying underwater paint, all contribute to the overall equation.

Not all these factors are controllable; for example, vessels still dry dock in unsatisfactory weather conditions, they are still anchored awaiting discharge in high temperature sea water areas and are always subject to possible mechanical damage to the underwater paint surfaces. It is possible nowadays to measure the hull roughness by means of a fairly cheap instrument. The proper use of this piece of equipment can help maintain hull roughness at acceptable levels when used during scheduled dry dockings.

5. Other performance improvers

There are other items which can be incorporated into a vessel when it is at the building stage, but these are not suitable candidates for retrofitting mainly because of the high cost.

A. Fuel oil storage tanks

Quite a large amount of heat is lost when storage tanks are mounted in direct contact with the sea. If double bottom tanks are involved little can be done, but if deep tanks are provided it is a comparatively simple matter to provide a cofferdam between side shell and the tank's outboard bulkhead. This works on a similar principle to that used for double glazing in that air is a poor conductor of heat.

B. Combined incinerator boiler

We will read in a later chapter about the deterioration of fuel quality. One side-effect is the generation of appreciable amounts of sludge recovered from the fuel, either in the settling tanks or from the purifier. This sludge does have a reasonable heat content and when suitably treated can be burnt in a combined incinerator/boiler, thus producing useful steam at a rather low cost. When burnt in a conventional incinerator the sludge has no useful contribution to the performance of the auxiliary plant. The sludge can be quite expensive to get rid of in certain parts of the world, and burning it in an incinerator/boiler can produce substantial savings.

C. Accommodation heating

It is possible to provide a link from the main cooling system to the accommodation heating system thus allowing the heat in the cooling system to heat the accommodation. This saves valuable steam which can be redirected to the steam alternator, if fitted, which in turn will produce more kW.

D. Two-speed pumps

When operating in cold sea water temperatures, only small quantities of sea water are necessary for cooling the various systems. The provision of two-speed motors on the sea-water cooling pumps represents a considerable energy saving when operating in the colder climates.

6. Summary of Chapter One

Some of the various technical possibilities for improving performance have been reviewed. There are many others which have not been mentioned and the author apologises to those who may feel slighted by these omissions.

Shipowners must decide which of the possibilities are the most suitable for their needs, bearing in mind that vessels on existing long-term charters already performing within the warranted performance need do nothing unless it is paid for by the charterer.

A sobering thought is that one of the most successful Panamax designs known to the author does not employ any of the sophisticated means for improving flow into the propeller, but relies simply on a good underwater hull geometry coupled with a simple barge type after body with Skeg (as shown in Fig. 5).

Fig. 5.   B & W Panamax aft end

CHAPTER TWO

Fuels

1. Crude oil

Most, if not all, of the oil in shipboard use is derived from crude oil; this includes heavy residual fuel used in the Main Engine, distillate or blended fuel used in the generators and even the lubricating oils and greases used to lubricate practically all machinery on board. The performance of a ship's machinery is directly affected by the quality of fuel oil, so various factors concerning its use will be examined in this chapter.

Crude oil is a naturally occurring substance found beneath the earth's crust in various parts of the world, notably the Middle East, USA, Russia and, more recently, the North Sea. It probably has its origins in decayed animal organisms from the wilderness of prehistoric times.

Contrary to popular belief, it is a comparatively light substance with an average specific gravity of around 0.8 kg/litre. On reflection, this is not surprising as it does contain petrol, kerosene and diesel oil as well as the heavy fuels and lubricants used aboard ship. The main components of crude oil are hydrogen and carbon which exist in various chemical compositions, depending on the geographical source of the crude.

In modern times, crude oil was first found in Pennsylvania in the 1850s and, after distillation into kerosene, was used as lamp oil. As industry expanded oil became the source of energy, first for steam boiler furnaces then for diesel engines during the early years of the twentieth century.

Quality and supply aspects remained fairly stable in the crude oil industry until 1973 when unnatural pressures were exerted to interrupt this stability, for mainly political reasons. The situation worsened progressively until the end of 1985, but since then there has been a move to return to some sort of stability in the supply and demand situation which, of course, affects the cost of shipboard fuels. It would, however, be very difficult to foresee the middle- and long-term prospects for the stability of crude oil prices.

It is unlikely that crude oil reserves will last as long as 50 years, so some sort of alternative fuel for ship propulsion will have to be found. Whether this alternative can compete with present-day crude oil prices will probably determine the medium- to long-term price of crude oil. Whereas there has been a recent encouraging downward trend in the price of crude oil which automatically affects bunker costs, there has been no such improvement in the quality of bunkers — nor is there any likelihood of this happening. The reason for this is partially on account of the 1973 crisis but also the oil industry's long-term economic strategy to obtain more product from the crude barrel. This inevitably has the effect of lowering the quality of residual fuel for shipboard use.

Fig. 6.   Refinery flow diagram

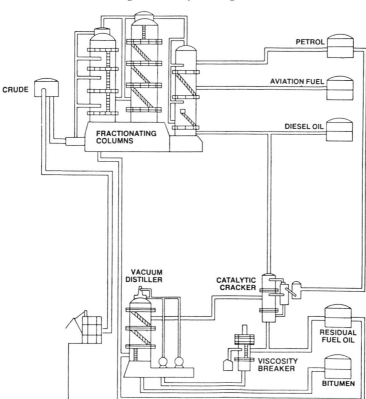

2. Refinery process

Crude oil is converted to distillate and residual fuel in the refinery process and a simple
flow diagram is shown in Fig. 6. In modern refineries secondary methods incorporating
catalytic crackers and viz breakers perform the function of obtaining more product from
the crude barrel. Certain areas, notably the Middle East, still have older refineries without
these expensive secondary units and fuel quality, as expressed by the specific gravity,
is rather good. In the natural order of things, these older refineries will eventually disappear
and fuel bunkers worldwide will probably have similar properties, particularly from a
specific gravity point of view.

The high quality of aviation and automobile fuels is sacrosanct and quite rightly so as
it would be quite dangerous to lower standards in these industries for obvious safety
reasons. This leaves only residual and so-called diesel oil as the targets of the oil industry's
goals.

A simple description of the refinery process begins when heated crude is fed into the
fractionating columns, in which vapourised products are led off at appropriate levels in
relation to their boiling temperatures. This primary stage deals with the light ends —
namely, petrol, kerosene and gas oil. Heavy products from the fractionating columns are
then fed into the vacuum distiller, which is the last chain in the primary process. The

products are then fed into the cracking or secondary section in which further light ends are extracted to the detriment of the residual components' quality. This is only a very simple explanation of what is a very complicated process and many more items of equipment are installed in a modern refinery. An illustration of a modern refinery complex is shown in Fig. 7.

Fig. 7.   Refinery complex

There are some modern refineries on stream which do not have any residual fuel left in the final process. There remains only petcoke which cannot presently be used for shipboard fuels.

It has been calculated that less than 5% of all crude oil is actually used as fuel aboard ships. This probably had an input in the oil majors' decision-making process for upgrading oil refineries to the detriment of this low-volume consumer's quality.

3. Residual fuel oil (heavy oil)

Heavy oil is a mixture mainly of hydrogen and carbon but also consisting of naturally occurring impurities such as sulphur, sodium, vanadium, water and ash. Its heat or energy content is directly related to its specific gravity, or density as it is nowadays called. The higher the density — the lower the heat value, for reasons now explained. Hydrogen has the higher heat value but lower density, conversely carbon has the lower heat value but the higher density. Carbon is the major constituent of fuel oil accounting for over 80% of the total. It follows that the higher the density the greater the carbon content and, therefore, less heat content.

Sulphur also has a rather low heat value but this constituent on average has only around 3% of the total content, but it should be taken into account when calculating the heat value. Apart from water, the other impurities do not affect the heat value to any significant extent.

Heavy oil when using primary refining processes had a density of around 0.96. Nowadays it is controlled to 0.991 to tie up with British Standard MA 100 1982. For many years the author was a member of the committee which formulated this standard which helped fill a void as there was no other suitable marine fuel standard in existence.

Actually, the density of the residual as delivered by a modern refinery is over 1.00. But, mainly because of present shipboard operational problems in using this residual, it is cut back to 0.991 before being sold as heavy fuel oil in the marine market — now predominantly diesel propelled. No restriction on gravity is necessary for steam powered plants, and the few steamers still remaining (also land based steam plants) use fuel of 1.00 density or even higher.

Modern purifying equipment can remove water from fuel oil even when it has a gravity above 1.00. So, in the short- to medium-term, fuel which has a gravity of around 1.01 will be increasingly available as vessels with this equipment enter service or existing vessels are retrofitted. Some oil majors offer discounts for using this fuel and an economic case for retrofitting could exist for certain vessels, depending on trading patterns and quantity of fuel used per day.

4. Distillate fuel (diesel and gas oil)

The quality of distillate fuel using primary refining processes used to be stabilised at a density perhaps of around 0.85 with no measurable contaminants contained in the delivered fuel. Gas oil quality marginally exceeded that of diesel oil, mainly on account of its superior combustibility properties necessary for very high speed diesel engines. Both these fuels were referred to as straight-run distillates. Secondary refining processes have almost eliminated these straight-run distillates from the marine market, although they are available for special purposes at a much higher cost.

Density of fuel, now delivered as diesel oil, is approaching 0.90 and certain contaminants are found. Even so, most diesel generators can burn this fuel without any problems. One could argue that diesel oil, as supplied from the old refineries was, in fact, of too good a quality in the vast majority of cases. A fact not overlooked by the oil majors when their future plans were being formulated.

5. Viscosity

Historically all marine fuels have been sold on a viscosity basis. The origins of this are not exactly clear, unless fuel was so cheap in the pre-1973 era that all other properties had no apparent contribution to its saleable value.

In Europe the unit used to measure viscosity was Redwood No. 1 at 100°F. In the 1970s this was changed to centistokes at 50°C for residual and at 40°C for so-called distillate fuels. It is due to be changed again in the near future and it will then use centistokes at 100°C as the basis for the measurement when an ISO standard is introduced shortly. What is now 380 cst fuel at 50°C will become 35 cst at 100°C and 180 cst at 50°C will become 25 cst at 100°C. For the benefits of readers a chart showing a draft of the new fuel type

Fig. 8.   Fuel classification

Requirements (June 1986) for heavy fuels for diesel engines (as bunkered)

Principal grades

Characteristic	Dim.	Limit	CIMAC A10	CIMAC B10	CIMAC C10	CIMAC D15	CIMAC E25	CIMAC F25	CIMAC G35	CIMAC H35	CIMAC K35	CIMAC H45	CIMAC K45	CIMAC H55	CIMAC K55
Designation — Draft ISO-F-			RMA10	RMB10	RMC10	RMD15	RME25	RMF25	RMG35	RMH35	RMK35	RMH45	RMK45	RMH55	—
Related to BS MA100 1982 (CIMAC 1982)				M4 3)	—	M5	—	M6	—	M7	—	M8	—	M9	—
Density at 15°C	kg/m³	max	975	991	991	991	991	991	991	991	1010	991	1010	991	1010
Kinematic viscosity at 100°C	cSt 1)	max	10	10	10	15	25	25	35	35	35	45	45	55	55
Flash point	°C	min	60	60	60	60	60	60	60	60	60	60	60	60	60
Pour point 2)	°C	max	0 / 6	24	24	30	30	30	30	30	30	30	30	30	30
Carbon Residue (Conradson)	%(m/m)	max	10	10	14	14	15	20	18	22	22	22	22	22	22
Ash	%(m/m)	max	0.10	0.10	0.10	0.10	0.10	0.15	0.15	0.20	0.20	0.20	0.20	0.20	0.20
Water	%(V/V)	max	0.50	0.50	0.50	0.80	1.0	1.0	1.0	1.0	1.0	1.0	1.0	1.0	1.0
Sulphur	%(m/m)	max	3.5	3.5	3.5	4.0	5.0	5.0	5.0	5.0	5.0	5.0	5.0	5.0	5.0
Vanadium	mg/kg	max	150	150	300	350	200	500	300	600	600	600	600	600	600
Aluminium	mg/kg	max	30	30	30	30	30	30	30	30	30	30	30	30	30
Total sediment after ageing		max	4)	4)	4)	4)	4)	4)	4)	4)	4)	4)	4)	4)	4)

1) Approximate equivalent viscosities (for information only):

Kinematic viscosity (cSt) at 100°C	10	15	25	35	45	55
Kinematic viscosity (cSt) at 80°C	15	25	45	75	100	130
Kinematic viscosity (cSt) at 50°C	40	80	180	380	500	700
Kinematic viscosity (cSt) at 40°C	14					

2) Where relevant: upper value winter quality, bottom value summer quality

3) Carbon Residue 12 for BS grade M4

4) No standard test method agreed. Fuel shall not cause excessive sludge

designations is shown in Figure 8. It is, however, expected that the market-place will continue to use the current practice of buying fuels with a 50°C IFO designation.

The viscosity of a fuel determines at what temperature it can be handled and burnt in the engine. It is a measurement of resistance to flow and increasing the temperature of a fuel reduces its viscosity as it becomes thinner or less viscous.

It is very important that fuel is burnt in a diesel engine at the engine builders' recommended viscosity. At normal temperature a typical fuel will have a viscosity of around 1,400 cst and this will fall to around 380 cst at 50°C and 15 cst at 130°C at which temperature it is usually burnt in most modern diesel engines. These modern engines are capable of burning fuel of up to 700 cst at 50°C which requires a temperature of around 140°C at the engine.

This fuel will gradually appear in the market-place as more modern vessels are delivered and distribution problems are solved by the oil majors. This very viscous fuel requires more heating requirements, both in the shore storage and distribution system also on board ship. These problems will gradually be overcome and in the medium-term 700 cst fuel will probably be commonplace.

For existing vessels it should be emphasised that only fuel with a viscosity falling within the installed heating capability should be used. However, it can be commercially attractive to upgrade certain vessels enabling high viscosities to be used.

Viscosity of distillate fuels is not normally a problem and is in the range of 4–10 cst at 40°C which is easily burnable without heating — except in Arctic-type conditions or in extreme known incidents when high-pour diesel oil was delivered.

6. Density

We have already touched upon typical figures for expected densities of modern fuels namely 0.991 and 0.900, respectively, for heavy and diesel oil — both of which are measured at 15°C.

If heavy fuel above 0.991 is supplied to vessels not equipped with the latest purification plant, water removal becomes impossible and main engine operational problems can be expected with reduced performance being one of the results. Most fuel is sold by weight but measured by volume in the tanks of the bunker supplier's barge or at his pipeline meter.

To convert volume to weight requires an accurate knowledge of the density at the delivered temperature which is usually converted to density at 15°C and then to weight. It is not unknown for unscrupulous suppliers to manipulate the density figures with obvious financial benefit to themselves. Very considerable sums of money can be involved, particularly in areas where low density fuels are still available and the delivery note density when manipulated higher by the supplier still does not approach that regarded as dangerous by the vessel's staff.

7. Compatibility

One of the most difficult properties to determine of a fuel is its compatibility, either inherent or when mixed with other fuels. It is unlikely that inherently incompatible fuel would be supplied to a vessel, but rare cases of this happening are on file. Mixing of fuels from different sources is a dangerous practice which can give rise to incompatibility and an important risk is eliminated if mixing of fuels in bunker tanks aboard ship is not allowed.

When using blended fuels in diesel generators a possibility exists of incompatibility problems arising and little can be done if the blended fuels are in fact incompatible. This, however, would not appear to be a major problem in practice.

A fuel is incompatible when it cannot hold the asphaltenes in solution. These come out of solution and form sludge which can cause catastrophic problems with the fuel system, leading to complete immobilisation in extreme cases.

There is a simple method of checking for compatibility called the spot test which involves diluting a sample of oil to be tested with a distillate oil and allowing a drop of the mixture to dry on a type of blotting paper. The dried spot is compared with a reference list of five spots having a rating of one to five. Rating one is good and incompatibility problems are unlikely. The spot test is not entirely reliable and a more reliable test is the sediment hot filtration test (SHF) carried out in a laboratory. This is more accurate and gives quantitative results rather than a sometimes difficult to read spot rating. The SHF test gives a reading which indicates the amount of sediment filtered out under certain specified conditions which is given as a percentage of the whole. A reading of 0.15% or above would give rise for concern as it would indicate incompatibility problems are likely.

A new test, which has not yet been introduced into the Institute of Petroleum test methods, is the Total Sediment (Potential) test. This new test incorporates a simulated ageing process and it will give an indication of the oil's behaviour whilst being treated which the SHF test did not always do.

8. Calorific value

The calorific value of a fuel is, in effect, its heat or energy content. When measuring the performance of an engine the results have to be referred back to the calorific value.

As mentioned earlier — the heavier the fuel, the lower the calorific value. Also mentioned earlier, fuel is sold on a viscosity basis which has no direct connection with energy. When discussing the performance of heavy oil and distillate fuels there is a difference between both viscosity and calorific value, but when unit cost is added to the equation the results may surprise some readers. Nowadays, calorific value is measured in megajoules per kilogram (mj/kg) whereas formerly the units of K.cals/kg or BTUs/lb were in general use. By adopting the unit cost approach to examine the energy performance of various fuels we get the following results (Table 3) based on actual examples (mid-1986 costs).

Table 3.   **Energy performance of various fuels**

Fuel type	Viscosity	Density	Calorific value	Cost/ tonne $	Unit cost
Heavy oil "A"	380	.991	40.00	45	1.125
Heavy oil "B"	180	.950	40.50	46	1.135
Diesel oil	10	.900	42.42	150	3.536
Straight-run distillate	5	.860	42.92	160	3.728

The extremely high unit cost of diesel oil and straight-run distillate will be noticed. It is for this reason that the use of blended fuel in diesel generators became popular. It will also be noticed that paying a $1 differential for 180 cst fuel is commercially not a good idea based on energy content alone. When fuel prices are higher this differential is normally greater and a good case for converting vessels to burn 380 cst fuel, or allowing those so fitted to use the heavier grade, could be made in suitable cases.

Using heavy oil in Main Engines only became generally adopted in the 1950s when the unit cost approach showed that even at the low fuel costs ruling then ($14 heavy and $27 diesel) it was commercially attractive to pay the high conversion costs.

9. Quality testing

It is frequently mentioned that the quality of fuels has deteriorated in recent years and we have attempted to qualify what this means in quantitative terms. Apart from those already mentioned, there are other aspects to consider. One of these is the presence of catalytic fines in the delivered bunkers, probably caused by bad housekeeping in the refinery.

Several organisations mainly associated with Classification Societies exist which will test fuel samples submitted to them and give results with recommendations within several days of them being sent from the vessel.

Catalytic fines are only one of the dangerous contaminants looked for. Should these abrasive fines enter the fuel system, rapid wear on all the metallic surfaces in contact will occur. Other contaminants found in fuel oil are calcium, iron and lead occasioned by dumping of used automobile lubricants into the refinery's crude supply. There have even been cases whereby chemicals have allegedly been dumped this way. In an extreme case naphthenic acid was allegedly the cause of extremely rapid wear on fuel pump plungers necessitating emergency repairs in a port of refuge. Normal quality testing would not identify contaminants such as this, but knowing the effect of the fuel on the fuel pumps gave certain leads which made identification that much easier.

Vanadium is a naturally occurring contaminant in the crude source which is not eliminated in the refinery process or the shipboard treatment plant. Excessive vanadium has a fouling effect in the steam generating sections of water tube boilers and, more importantly, on the exhaust valves of four stroke engines where it deposits itself to the detriment of the valves' efficiency.

Sodium can occur naturally in the crude supply or it can be introduced by contamination during its passage through the distribution network. If not removed by the purification plant, it can cause fouling of the turbochargers and gas passages within the Main Engine and on the heating surfaces of the economiser. The level of Conradson carbon and asphaltene is also given which are indicators of the combustibility of the fuel. Several cases in which fuel pump plungers seized were attributed to a higher than usual asphaltene content. Another undesirable feature is the presence of wax which leads to high pour point fuels, and this can cause problems transferring and handling the fuel if temperatures are allowed to fall close to the pour point and the wax solidifies.

All these and several other contaminants are looked for in each sample. Additionally, a check on the density and viscosity is given to ensure the specification is met. We must not forget water which is often found in fuel oils and can reach up to 1%, the recognised commercial limit.

Fig. 9   Fuel oil test kit

Fig. 10   Bunker delivery note

INTERNATIONAL CHAMBER OF SHIPPING
RECOMMENDED BUNKER DELIVERY NOTE

1. Product name or general designation	
2. Quantity in tonnes	
3. Viscosity (cSt) at 50°C (residual fuels) or 40°C (distillate fuels)	
4. Relative Density at 15°C	
5. Flash Point (P-M) Closed Cup (°C)	
6. Water Content Volume (percentage)	
7. Pour Point (°C)	
8. Conradson Carbon Residue (percentage weight)	
9. Sediment (percentage weight)	
10. Vanadium (ppm)	
11. Sulphur (percentage weight)	
12. Cetane Number (distillate fuels only)	

Note: Specific energy (calorific value) (MJ/kg) can be calculated from the known density and sulphur.

There are test kits suitable for shipboard use on the market. These, naturally, do not give the full range of tests available from the shore-based organisations but can be used as a screening device to decide if sufficient evidence exists to send the fuel to the experts for a full analysis. The test kit most familiar to the author is shown in Fig. 9. The tests incorporated in this kit are:

1. Density.
2. Viscosity.
3. Water content (fresh or salt).
4. Compatibility.
5. Conradson carbon.

The oil majors have quite tight contracts when it comes to acceptance of their delivered quantities, but a quick test on density with a shipboard kit could avoid lengthy discussions later. Samples must be taken at each bunkering, whether or not it is intended to send them for analysis or test them aboard. Ideally, the samples should be made by a controlled drip to ensure they are representative and should be held aboard for several months. Not only will this action help prove if the fuel is unsuitable in the event of problems developing, but it will help defend any claim made for pollution under MARPOL regulations.

10. Documentation

Considering the amount of money involved in the purchase of fuel, very little documentation exists. In a move to regularise the situation, the International Chamber of Shipping (ICS) developed a Bunker Delivery Note in 1979 which contained all the information necessary to enable the shipowner to decide if what he had received was satisfactory and was as had been ordered. This delivery note, a copy of which is shown in Fig. 10, has not proved to be a success. Only a concerted effort by shipowners and their representative organisations will make this useful document internationally acceptable.

ICS in conjunction with BIMCO drew up a Bunker Quality Control clause some time ago in an attempt to apportion responsibility in the event of bunkers giving unsatisfactory performance being supplied. Since this clause was introduced, the market has been decidedly poor, which makes it difficult for shipowners to have this clause accepted by charterers. The author was a member of the committee which formulated these documents and it is sad to see all the efforts that were made being unacceptable due to current market conditions. An extract from the BIMCO clause is appended:

BIMCO Bunker Quality Control Clause for Time Charters

(a) The Charterers shall supply bunkers of a quality suitable for burning in the Vessel's engines and auxiliaries and which conform to the specifications mutually agreed under this charter.

(b) At the time of delivery of the Vessel the Owners shall place at the disposal of the Charterers, the bunker delivery notes and any samples relating to the fuels existing onboard.

(c) During the currency of the charter, the Charterers shall ensure that delivery notes are presented to the Vessel on the delivery of fuels and that during bunkering, representative samples of the fuels supplied shall be taken at the Vessel's bunkering manifold and sealed in the presence of competent representatives of the Charterers and the Vessel.

(d) The fuel samples shall be retained by the Charterers and the Vessel for 90 (ninety) days after the date of delivery or for whatever period necessary in the case of a prior dispute and any dispute as to whether the bunker fuels conform to the agreed specifications shall be settled by analysis of the samples by (. . .) or by another mutually agreed fuels analyst whose findings shall be conclusive evidence as to conformity or otherwise with the bunker fuels specifications.

(e) The Owners reserve their right to make a claim against the Charterers for any damage to the Main Engines or the auxiliaries caused by the use of unsuitable fuels or fuel not complying with the agreed specifications. Additionally, if bunker fuels supplied do not conform with the mutually agreed specifications or otherwise prove unsuitable for burning in the ship's engines or auxiliaries, the Owners shall not be held responsible for any reduction in the Vessel's speed performance and/or increased bunker consumption, nor for any time lost and any other consequences.

11. Combustibility

To define a heavy fuel oil's combustibility or ignition quality is rather difficult, as no internationally approved method is in existence. This is not the case for distillate fuels for which a cetane number is available. This gives a very good idea of the fuel's combustibility under test running conditions. Some oil majors have attempted to rectify this situation and Shell have introduced what they call the Calculated Carbon Aromacity Index (CCAI) which appears to have success when defining ignition quality of residual fuels.

CCAI is given as a non-dimensional number in the range 750-975. Below 850 is considered to be a very good reading, probably giving troublefree combustion, whereas above 875 is considered to be a poor reading, giving rise to poor combustion. As an example of its use, we could choose grade M7 from British Standard MA100 with a density of 0.991 and a viscosity of 380 cst at 50°C. This gives a CCAI of around 852 more or less in the middle of the range. Even the fuel of the future with a 1.01 density and a viscosity of 700 cst at 50°C gives a CCAI of 865.

Returning to the first example — namely grade M7 — if we substitute 180 cst at 50°C for 380 cst at 50°C we get a CCAI of around 860, a somewhat worse reading for what might appear to be a better fuel on paper.

Certain rather old four-stroke engines experience combustion problems when operating on modern fuels. In an actual example, fuel of 80 cst at 50°C with a density of 0.980 gave problems, even though the CCAI was 860. The specification was changed to 80 cst at 50°C with a density of 0.960 and no further trouble was experienced; the CCAI of this new fuel being 840.

It would appear that some engines do have a CCAI threshold limit, but published data is hard to come by.

A nomograph produced by Shell is shown in Figure 11 for the benefit of readers. Shell also have a programmable calculator which gives all manner of functions which can assist in evaluating bunker quality. These include:

- Specific energy
- Viscosity
- CCAI
- Blending ratio
- Density/temperature corrections

Fig. 11   Nomograph for determining CCAI

NOMOGRAPH
CALCULATED CARBON AROMATICITY INDEX
Based on equation CCAI = D − 141 loglog (V + 0.85) − 81

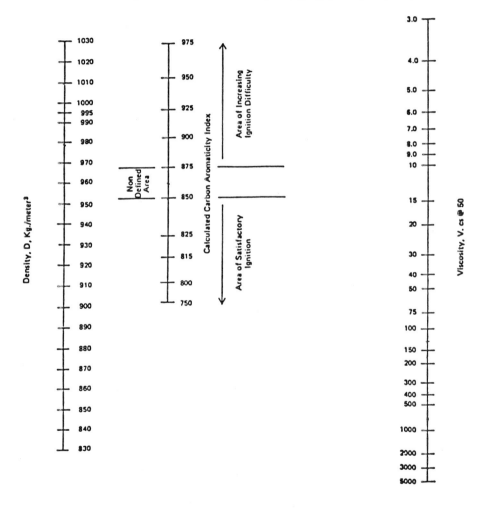

12. Fuel additives

The use of chemical additives in fuels has been commonplace for many years. Its origin probably goes back to pre-1950 when survey of oil fuel storage tanks at every special survey was a classification requirement. Chemicals were then used to reduce or eliminate the amount of sludge left in these tanks, thus reducing cleaning costs and off-hire time. These very inconvenient surveys have now lapsed, but chemical additives received a new lease of life when fuel quality started deteriorating in the 1970s.

When chemicals are used for sludge-breaking purposes, before the combustion process has taken place, it would appear that they have a certain amount of success. They would also appear to have a contribution to make in improving the scavenge space cleanliness on certain loop scavenged engines. Their effectiveness as a combustion improver on an energy content basis is open to discussion. To conduct a shop test, in an attempt to prove this aspect, would be very costly in the case of even a moderately powered engine. This course of action would appear to be the best solution for those wishing to pursue the claims of their products which could then be backed up by some sort of guarantee.

Water has also been introduced into fuel in a controlled amount and then emulsified before injection into the cylinders. Various shop tests have been conducted, albeit on smaller engine types than those normally associated with those in this review. Claims of fuel consumption improvements of around 3% have been made; also claims of improvement in engine cleanliness and reduction in N.Ox (nitrogen oxide) emission. One problem which could be expected when using this emulsified fuel is in the slow steaming mode, when extremely low exhaust gas exit temperatures occur and dew point corrosion problems rear their ugly head.

13. Quantity audit

Several organisations will offer to carry out a bunker audit to check if the vessel is receiving the amount of fuel shown on the receipt note. The surveyors carrying out this task are very experienced in this type of work and have the advantage of being well trained and are supported by a specialised organisation at their head office. The author's experience is that the money recovered from the bunker supplier for shortage in supply matched the fee charged by the audit company. When bunkering in areas where serious shortage problems have been experienced, the occasional use of a bunker audit could prove to be a deterrent.

14. Cost control

Depending, of course, on unit cost, fuel oil represents a major part of most vessels' operating costs. When heavy oil approached $200 per tonne, fuel costs accounted for over 50% of most vessels' operating budgets. At $80 per tonne it probably equals the total manning cost on our typical vessel even when slow steaming.

Bunker purchasing is a specialised subject requiring very experienced personnel who ideally should be aware both of the technical aspects discussed in this chapter as well as the purely commercial aspects. The unit cost of energy should figure in any calculations made in determining the favoured supplier. Even credit terms should be taken into account when making a selection. Associated charges, such as for barging, are not always indicated when suppliers make quotations and due allowance should be made for comparison purposes.

The strategy of deciding the amount and supply location of bunkers is of utmost importance for spot voyage charters. A classical example is a Panamax vessel loading New Orleans for Japan and requiring bunkers before arrival at the discharge port. The main inputs to the equation are costs of fuel at loading port, fuel costs at a West Coast, USA, port and the freight rate.

If freight rates are high, it is normally more attractive to maximise cargo deadweight using maximum available draught through the canal, and pick up bunkers at Long Beach where costs are usually very competitive.

If freight rates are low and the differential in fuel costs between New Orleans and Long Beach is greater in Long Beach's favour than the freight rate, it is also more attractive to bunker at Long Beach. The only attraction in bunkering at New Orleans or another bunker port before the canal would be if freight rates were low and fuel cost difference in New Orleans favour greater than the freight rate. The difference in sailing distance by calling at Long Beach for bunkers is negligible if using a great circle route.

Should the vessel be coming off or on hire in a high cost area, for example, Japan, it is commercially attractive for the shipowner to fill her with bunkers in a low cost port such as Long Beach to obtain the Japanese fuel costs if this is allowed in the charterparty. Conversely if a vessel is coming off or on hire in a low fuel cost port it would pay to sail from a high cost port with minimum allowable bunkers on board.

Ship's staff should be encouraged to give accurate bunker requirements when requested to submit figures. Not only does this mean that optimised quantities at beneficial costs can be taken advantage of but also that any fuel turned away because the ship's tanks are full, does not incur a cost penalty from the supplier. Alternatively, if insufficient fuel has been ordered, a penalty may be incurred by the barge having to make a return trip to make up the shortfall.

Regarding the actual cost of bunkers *Lloyd's List* gives a weekly rundown on world-wide costs which are an extremely useful guide. There are many specialised firms who will buy your bunkers for you based on their vast experience and commercial clout.

15. Summary of Chapter Two

Fuels costs have at last paused in their ever-increasing upwards spiral. Whether this is temporary or not remains to be seen. We will have to find an alternative energy source in the medium- to long-term, although the emphasis on doing this will no doubt subside due to the recent cost reductions. Shipowners will have to prepare themselves for the quality of fuel oil expected in the short- to medium-term. It is hoped that this chapter will have alerted them to the various issues involved.

CHAPTER THREE

Performance monitoring

1. Measuring methods used

Performance monitoring systems can extend to very comprehensive systems using sophisticated equipment such as Doppler logs and cathode ray tubes. Questions that should be asked by any prospective user are: what is expected of the system, will it be cost-effective and do the people entrusted with its use fully understand the basic principles involved?

Probably the most important parameter is the ship's speed, both over the ground and through the water. Most ships now have Satnav, so speed over the ground is available on demand or at least when a satellite is in a suitable position. Speed through the water is a little more difficult and a Doppler log gives very accurate results at high capital cost. Cheaper electric magnetic or venturi type logs are available but are not as accurate as the Doppler log.

Next in importance is the developed horsepower of the main propulsion system. The classical method is to take indicator cards from each cylinder and calculate the mean indicated pressure and thence the indicated horsepower using a Planimeter. When used by an expert, a Planimeter is fairly accurate but suffers from the effects of any shipboard vibration and requires a steady hand. The mechanical efficiency of the engine has to be known so that brake horsepower can be calculated and the whole operation, perhaps, has a 5% measurable tolerance.

Torsionmeters are available for measuring shaft horsepower and can be coupled with a strain gauge for measuring thrust. Accuracy within a 2% tolerance is a reasonable target for this type of equipment. By replacing the indicator with a cathode ray tube, it is possible to obtain a more accurate measurement of the indicated horsepower, possibly within a 2% tolerance.

Main Engine revolutions are accurately measured on most vessels using comparatively simple equipment. This parameter is probably the easiest of all to measure and is very useful in determining trends, as will be discussed later.

Next we come to fuel consumption which, being directly related to horsepower, is rather important and in the case of the performance of chartered vessel's fuel consumption and speed are the two most used parameters in determining if a vessel is performing to charterparty warranties or not.

Fuel consumption can be measured by flowmeters or simply by tank soundings and although this latter method is subject to large daily fluctuations it is reasonably accurate over a period. In the case of charterparty off-hire surveys tank soundings are still the only method used. Using flowmeters has its uses, particularly when Main Engine performance is being monitored, but they are of only limited use for charterparty purposes. A point to remember is that in the case of modern fuels usually containing amounts of sludge and water, the flowmeter does not measure these contaminants, so using a flowmeter only would result in a deficit when trying to balance fuel consumed with the closing stock.

The measurement of weather is important as this arguably has the greatest impact on a vessel's performance. Apart from wind speed and direction instruments (ananometers), very little instrumentation exists and reliance on visual observations is the norm.

Various pressures and temperatures in the propulsion machinery systems are also measured using standard thermometers, pyrometers and pressure gauges to enable laid down limits to be observed.

These then are the parameters requiring measurement and the means generally used aboard ship to accomplish this. Some comments on the practical aspects of each follow.

2. Ship's speed

The movement of a ship through the water is subject to many actions. First, we have the frictional resistance of the hull surfaces and the resistance imposed by wave and eddy making. The sea itself is far from being stationary and ocean currents as well as swell must be taken into account. Even air resistance has to be considered, especially on high-sided car carriers, container vessels and the like.

Using Satnav can give a reliable speed over the ground which is satisfactory for practical purposes. Most charterparty warranties use speed over the ground and sea trial speed runs also measure speed over the ground. Measurement of speed through the water requires a more technical approach, and if no extraneous forces were acting on the hull, speed through the water would be the same as over the ground for all practical purposes. By comparing speed through the water with speed over the ground we can calculate the effect of ocean currents on the vessel. We would still be unable to decide the strength and direction of the current only its resultant effect.

Most ships are equipped with Ocean Routeing Charts from which the prevailing current speed and direction and also the wind force/direction can be obtained. These are based on historical data and are in a monthly form covering the most frequently used trade routes. There are also the weather services to turn to if up-to-date weather patterns are required for either present voyage planning or past performance analysis.

It will be seen that there are various options for determining if speed over the ground is being affected by extraneous forces. If this is not the case, and the ship's speed is down, then it can be assumed in most instances to be due to hull fouling.

3. Horsepower

Horsepower is transmitted to the propeller from the Main Engine cylinders via pistons, connecting rods, crankshaft and, finally, the propeller shaft. Horsepower is normally measured by means of an indicator or cathode ray tube which measures pressure within the cylinders over a full working cycle. The resulting pressure volume diagram is converted first to mean indicated pressure and then indicated horsepower.

This is the power developed in the cylinders and it must be converted to brake horsepower, which is the indicated horsepower less the frictional losses in the transmission system — notably piston rings against liner wall. Some modern engines have actually reduced the number of piston rings to increase the mechanical efficiency, which is the same as reducing frictional losses. Mechanical efficiencies of around 93% are claimed for these modern engines.

To be strictly correct, the propeller horsepower should be used in hydrodynamic calculations. These take into account the losses between the brake, normally taken at the flywheel and the propeller. Frictional resistance in the shaft bearings, stern tube bearing and sealing glands are included in this calculation; these probably amount to around 1½%, which must be deducted from the brake horsepower.

Torsionmeters are sometimes used to measure horsepower and work on the principle of the degree of twisting in the intermediate shaft being proportional to the torque. When the measured torque is multiplied by the RPM, the brake horsepower can be derived.

4. Revolutions/slip

Revolutions are indicated in analogue form by means of a tachometer, which gives a visual indication of the revolutions being turned. The average revolutions shown in the abstracts and logs are measured by an integrating counter which gives rather accurate results.

The engine speed is derived from multiplying the average revolutions by a propeller constant. This constant is calculated by the propeller designer and is obtained from the mean face pitch. When multiplied by the average revolutions this gives the theoretical advance of the vessel, and when compared with the vessel's speed over the ground gives the apparent slip as:

$$\frac{E - S}{E}$$

where E is the engine speed and S the ship's speed over the ground.

Hydrodynamicists generally use the true slip in their calculations and, in the above formulae, S becomes the ship's speed minus the wake speed. For practical purposes wake speed is not measurable with any degree of accuracy, and apparent slip is favoured in this study.

Actually, the calculation of mean face pitch can give rise to some discussion. From the author's experience, vessels built in the Far East generally have lower apparent slips than European-built vessels when measured on sea trials. This is not to say they are more efficient, but different methods of calculating mean face pitch are presumably used. The important thing is that apparent slip is used for detecting trends rather than for vessel comparison purposes. Apparent slip has a very important input to the bottom line in a ship performance audit. A 1% change in slip is equal to a 3% change in fuel consumption, based on cube law considerations.

5. Fuel consumption

On a typical vessel there are usually three main consumers of fuel, namely, main propulsion, boilers and diesel generators. In some installations, the main propulsion and boiler fuel comes from the same service tank which can lead to measurement problems if this is not apportioned correctly.

Measurement of fuel flow to the main propulsion system is usually by flowmeter. Due to the complexity of the modern fuel system, placing of the flowmeter is rather important otherwise the amount of fuel in circulation will be measured. This is around three times the actual fuel consumed and, if this were measured, it would be meaningless.

We will trace the path of the fuel from bunker barge to engine so that the practical problems in accurately measuring fuel consumption are realised. The fuel being bunkered passes through the supplier's meter or by checking the soundings of his barge, and the quantity agreed by supplier and ship's staff. In some cases agreement cannot be reached but, in general, it would appear that, providing ship's staff take care and, more important, show interest, no marked differences occur between supplier's and the ship's figures.

Delivered fuel is then directed to various storage tanks around the vessel, usually double bottom or deep tanks. Ship's staff take soundings to check if the delivered quantity compares with the supplier's figures. Problems immediately arise because of various factors. For example, if double bottom storage tanks are being used then soundings are subject to large corrections in trim effect, which may not be easy accurately to determine. The temperature of the delivered fuel may well be known and can be easily measured, but its temperature in the double bottom tank can only be estimated. This would also apply to the density if it has been mixed with other fuels, although mixing fuels is a practice now frowned upon by most shipowners.

At sea, the fuel in the storage tanks is pumped to the settling tank and allowed to stand to enable water and sludge to settle out. These are drained off and a loss from the delivered quantity immediately arises depending, of course, on the amount drained off. The fuel is then fed into the service tank via the purification system, in which further amounts of water and impurities are removed by the extremely high gravity forces in the centrifugal purifiers.

The fuel is then fed into the engine via a flowmeter, if fitted. A certain amount of fuel leaks from parts of the system, notably fuel pumps and injectors; the amount of leakage is dependent on the condition of these delicate parts. This leakage must be deducted from the reading taken from the flowmeter if we are conducting a performance test.

The alternative method of measuring fuel consumption is simply by tank soundings which may be by hand-held tapes or by pressure transducers with a direct reading to a gauge. Experienced ship's staff will be able to determine fuel consumption reasonably accurately using this equipment over a voyage, but for accurate daily consumption intended for performance analysis a flowmeter is obviously more accurate.

When time-chartered vessels are subjected to on-hire and off-hire surveys the moment of truth arrives when independent surveyors check the quantity of fuel remaining aboard. Ideally this should check out with the amount of fuel remaining in the abstracts which, presumably, will have been analysed on the assumption that the daily fuel consumptions shown are in fact correct. However, some quite startling differences have been noted when off- and on-hire survey results have been compared with remains shown on the abstracts. Historically this can be traced back to the now discouraged practice of the chief engineer having fuel up his sleeve.

An exercise was recently undertaken by the author spanning a 12-month period covering 16 time-chartered vessels. The result showed a total book difference of around 325 tonnes or 0.057 tonnes per day per ship which would exceed the accuracy of most flowmeters.

We could say that over a period fuel consumption by tank soundings gives reasonable accuracy but, for performance monitoring purposes, a flowmeter would obviously be more suitable.

Ship Performance

6. Weather

When weather is mentioned, most seafarers immediately think of wind force as expressed by the Beaufort Scale, which for reference purposes is given hereunder:

Table 4. **Beaufort Scale**

Number	Description	Wind speed — knots
1	Light Air	2
2	Light Breeze	5
3	Gentle Breeze	9
4	Moderate Breeze	13
5	Fresh Breeze	19
6	Strong Breeze	24
7	Near Gale	30
8	Gale	37
9	Strong Gale	44
10	Storm	52
11	Violent Storm	60
12	Hurricane	Above 64

We must also consider in which direction the wind is blowing as obviously a wind force 4 on the bow will have a completely different effect on a vessel to a wind force 4 on the stern. For some unknown reason, standard charterparty weather conditions generally refer to the wind force only; not the direction.

Another weather factor affecting vessel performance is swell, which can be expressed by the following scale represented in order of magnitude:

Table 5. **Swell represented in order of magnitude**

		Intensity		
		Low	Mod	Heavy
Length	Short	2	4	7
with	Mod.	2	5	8
respect	Long	3	6	9
to ship's				
length	No swell		1	
No data	0		Very heavy swell 10	

Ocean currents are another phenomenon affecting vessel performance, and these can readily be expressed as their speed in knots and direction.

To simplify the recording of these aforementioned parameters we can use a simple sector system for relating the relative direction of either wind force, swell or current to the ship's heading as in Figure 12.

Fig. 12. **Simple sector system**

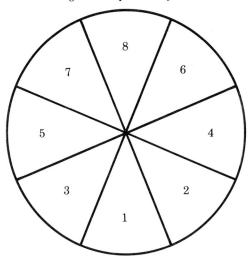

The log abstract or passage summary could then be simplified to a six figure entry as shown:

Table 6. Log abstract or passage summary

Wind		Swell		Current	
Sector	Force	Sector	Force	Sector	Speed

This will cover most of the weather conditions normally assumed to affect performance.

We can also introduce factors for determining what effect the weather is having on the vessel by adding a rating for pitching and rolling, also deck wetness. These important factors can be represented numerically in order of magnitude as follows:

Pitch/roll
1. None.
2. Slightly.
3. Easily.
4. Moderately at times.
5. Moderately.
6. Heavily at times.
7. Heavily.
8. Violently at times.
9. Violently.

Fore deck wetness
1. Dry.
2. Spray.
3. Wet.
4. Very wet.

All these criteria can now be recorded by a simple digital system on a daily basis as shown:

Table 7.   **Simple digital system for daily recording**

Date	Wind		Swell		Current		Pitch/ roll	Deck wetness
	Sector	Force	Sector	Force	Sector	Speed		
1.10.86	2	4	2	2	7	2	1	1
2.10.86	2	5	2	3	6	2	2	2
3.10.86	3	4	3	2	6	2	1	1
4.10.86	2	6	2	4	6	2	3	1

This represents only part of a typical voyage abstract or passage summary. Other data normally included such as RPM, position, course and various fuel consumptions will also be shown.

Although an improvement on the limited information normally given on a typical voyage abstract, it still has a disadvantage as it must rely on the accuracy of whoever is entrusted to enter the data. This arises when a weather pattern covering a 24-hour period has to be condensed into a single line entry. It is possible, of course, to split the day into watches in a similar manner as the deck and engine logs are filled in. This would be considered a retrograde step; abstracts were originally introduced to avoid the tediousness of examining the submitted logs. To return to this labour-intensive practice knowing that shore personnel are subjected to manning reductions in similar manner as their seagoing colleagues would not be popular.

Unless this proposed splitting of the day were confined to two distinct 12-hour periods, it is unlikely that it would be accepted under NYPE time charters for calculating heavy weather periods. Although it could possibly be used in some tanker charters at lesser periods than 12 hours.

Experience has shown that given proper instructions it is possible for seagoing personnel satisfactorily to enter data on a once per 24-hour period which will adequately describe average weather conditions and its effect on the vessel.

7. Pressures — temperatures

Various critical pressures and temperatures affect the performance of the vessel, starting with atmospheric pressure and the ambient temperatures of sea water and outside air. It is normal to convert these to ISO standards when comparing actual results to those recorded at the shop or sea trials. The main effect of pressures and temperatures is, of course, on the main propulsion machinery, although it is important that auxiliary machinery and steam plant are given due consideration.

Pressures and temperatures in the Main Engine combustion cycle are the most important means of measuring the efficiency of the plant. The theoretical considerations governing ideal cycle efficiency are that heat is supplied at a single high temperature, $T1$, and rejected at a single low temperature, $T2$. It can be shown that maximum efficiency is achieved when $T1$ is as high as possible and $T2$ is as low as possible in the form

$$\frac{T1 - T2}{T1}$$

In an existing plant we can achieve this by keeping the maximum cycle pressure (P.Max) to the shop trial figure and also keep the scavenge pressure, hence temperature, to shop trial figures having regard to ISO corrections. The mean effective pressure (P.Mean) is that converted directly to the output in horsepower and must be kept as high as possible for obvious reasons. The ratio between P.Max and P.Mean is a more practical means of determining cycle efficiency, as it is rather easy to measure pressures but rather difficult to measure cyclic temperatures. The higher this ratio, the higher the thermodynamic efficiency.

Another pressure used to measure plant efficiency is the compression pressure (P.Comp) this is the maximum pressure in the cylinder when the fuel is shut off; a comparatively easy measurement in practice. It can even be determined with the fuel kept on if accurate draw cards are taken with the PV indicator.

The relationship between P.Comp and scavenge pressure (P.Scav) gives a good guide to the condition of piston rings and amount of liner wear, which is, in effect, the gas tightness of the cylinder/piston configuration. By using this relationship we can monitor the condition of piston rings and liner whilst the engine is running, so that corrective action can be taken in port. Each engine type has its own P.Comp/P.Scav relationship and this is normally given in the instruction book.

Correct fuel injection timing is a prerequisite of optimised engine performance, and incorrect timing will almost certainly reduce the P.Max with resultant loss of efficiency. Incorrect timing can also be determined with the use of the PV diagram or the cathode ray tube.

Each fuel will have its own combustion properties and, in an ideal world, the fuel injection timing should be checked and adjusted each time bunkers are taken. Modern engines are arranged so that injection timing can be adjusted when the engine is running which makes this operation relatively simple. Different engines have their own mechanical or electronic means of accomplishing this and the equipment is nowadays known as variable injection timing (VIT).

So the difference between modern engines and older types is that the means of adjusting fuel injection timing is now done rather easily whereas it used to be a time-consuming task necessitating altering fuel pump cam position and/or adding/removing liners from the fuel pump housings.

This, more or less, covers the essential pressures in the combustion chamber which leaves us with some of the more important temperatures to consider.

One of the most important is the temperature of the fuel as it is injected into the cylinders. Actually, it is the viscosity we are really concerned about and it is normal nowadays to provide viscosity controllers, usually called viscotherms, to monitor this function. Viscotherms avoid the necessity of operators converting the viscosity of whatever fuel is in use to temperature. Bunker receipt notes are notorious for describing the viscosity of the fuel as it was ordered and, in practice, it is generally less. For example, 380 cst at 50°C may turn out to be 320 cst at 50°C.

There is also the problem of using fuel of different viscosities; the result of which may be difficult to determine. The viscotherm takes care of these difficulties and heats the fuel to enable the preset viscosity to be maintained.

Jacket water temperatures are now kept much higher than they used to be, nominally 85°C against the 65°C previously used. This increases thermal efficiency slightly and has the added function of reducing acidic attack on the cylinder liners of modern long stroke engines.

Exhaust gas temperatures are also important. In the case of cylinder outlet temperatures these represent T2 in the ideal cycle referred to earlier. In the case of impulse turbo-charged engines, cylinder outlet temperatures did not have any significant contribution except, perhaps, for determining trends. Constant pressure turbo-charged engines have a more discernible contribution and differences in individual cylinder outlets can indicate unequal power distribution which should be verified by indicator diagrams. Exhaust gas temperatures vary with the engine room ambient temperature and the scavenge air temperature. Any increase in these will manifest itself in increased exhaust gas temperatures. To ensure that the correct amount of combustion air is delivered to the cylinders, the scavenge temperature should be kept at the engine builders' recommended figures, so optimising efficiency.

However, by increasing the scavenge air temperature the exhaust gas temperature is also increased, and heat recovery from the economiser likewise increased. This should be borne in mind on plants having large exhaust gas recovery systems incorporating a turbo-alternator. At certain Main Engine loadings, a slight increase in scavenge temperature can mean an extra $20-30$kW output from the turbo-alternator.

8. Monitoring systems

So how do we monitor all these numerous measurements and present them in some sort of simple understandable form?

There are very sophisticated Main Engine monitoring systems already on the market and more are being developed. As this book is aimed at a representative readership, it is not intended to discuss fully these highly technical systems beyond saying that most of the basic considerations have been explained in this chapter. It might interest readers to know that some of these sophisticated systems are more technologically advanced than those used on most engine builders' test bays where, unlike ship's staff, most participants have achieved high academic standards.

The author has developed a rather simple manual system which is illustrated in Figure 13. It has been dubbed a manual system to distinguish it from a computerised system, which is the subject of a later chapter.

The vessel shown is a 1982-built Panamax with a rather conventional hull form, improved vessels with a superior engine performance and far better hulls are, of course, now available. RPM has been chosen as the base, mainly because this is probably the most accurate of all the many variables included in the study. Starting at the top curve on the graph and working downwards we start with P.Max as recorded at the sea trial when running on heavy oil. This should give a reasonable comparison against service results taken in good weather and, if too far adrift from the sea trial figure, further investigation is called for. As engines age and fuel oil quality deteriorate P.Max has a tendency to drop, and it should be maintained at the recommended level, always providing the engine condition is satisfactory.

The next set of curves represents the loading on the Main Engine caused mainly by the resistance of the vessel through the water and partially by the ruling weather conditions. Normal rating of the Main Engine as tested at the shop trial is the basis of the 100% curve. At sea trials it is normal for service power to be developed at around 103% service revolutions, so that a margin for future increased resistance is built into the propulsion system.

As resistance increases in service, so the absorbed load on the Main Engine increases irrespective of the actual output. To avoid problems identifying the output in terms of

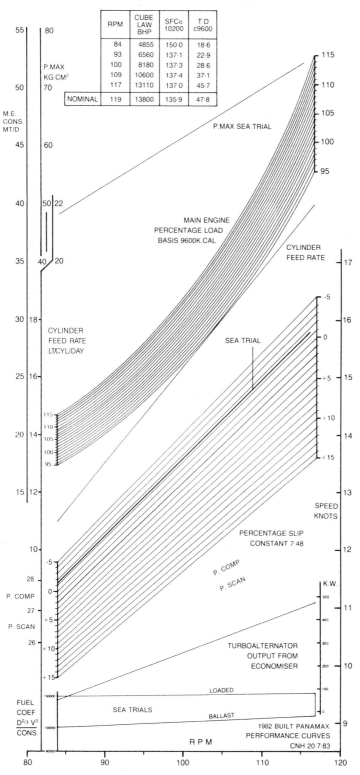

	RPM	CUBE LAW BHP	SFCe 10200	T·D c9600
	84	4855	150·0	18·6
	93	6560	137·1	22·9
	100	8180	137·3	28·6
	109	10600	137·4	37·1
	117	13110	137·0	45·7
NOMINAL	119	13800	135·9	47·8

P.MAX KG·CM²

M.E. CONS MT/D

P.MAX SEA TRIAL

MAIN ENGINE
PERCENTAGE LOAD
BASIS 9600K.CAL

CYLINDER
FEED RATE

CYLINDER
FEED RATE
LT/CYL/DAY

SEA TRIAL

SPEED
KNOTS

PERCENTAGE SLIP
CONSTANT 7·48

P. COMP

P. COMP
P. SCAN

P. SCAN

K.W.

TURBOALTERNATOR
OUTPUT FROM
ECONOMISER

FUEL
COEF
$\dfrac{D^{2/3}\ V^3}{CONS.}$

LOADED

SEA TRIALS

BALLAST

1982 BUILT PANAMAX
PERFORMANCE CURVES
CNH 20·7·83

R P M

80   90   100   110   120

Fig. 13.   Manual ship performance monitoring system

horsepower, fuel consumption has been used and the means of converting RPM to horsepower thence consumption is shown in the table.

The loading curves may not be accurate in absolute terms but they are very useful in determining trends, especially when operating in sea areas promoting underwater growth. Once the absorbed load rises above 115% a careful watch should be instituted until the vessel can be scrubbed or dry docked. A word of caution when using these curves with derated Main Engines, unless the derated horsepower and revolutions are used the results will show that in all probability overloading is taking place even though this is not strictly true.

The next curve shows cylinder lubricating oil feed rate which gives guidance on the consumption of this rather expensive commodity. It should not be used in isolation, but other factors must be considered. For example, cylinder wear rate and physical condition of the scavenge spaces with respect to its oily or dry appearance.

Curves showing apparent slip percentage are directly under the cylinder feed rate curve. These are useful for giving an instant slip percentage reading without the trouble of carrying out the calculation. They are very useful for determining trends and also for predicting future performance.

Next shown is the ratio in absolute pressures of P.Comp/P.Scav which, as mentioned previously, is a measure of the gas tightness of the piston ring-cylinder liner wall configuration; any fall off should receive further investigation.

The vessel depicted in the example is fitted with a steam turbo-alternator and the next curve shows the expected kW output over the range of revolutions. Should this be below expectations, further investigation is called for. For example, the dump steam valve may be open or excessive steam is being diverted to other less essential consumers.

Finally, the fuel coefficient — as calculated on the sea trials — is shown, which, of course, were in rather good weather. Due allowance must be made for weather conditions when service results are being compared with these sea trials results.

9. Speed versus consumption

The most important consideration of all performance aspects is the vessel's speed against consumption, both in laden and ballast conditions. All other parameters mentioned in this chapter are simply vehicles used in arriving at what might be called the bottom line of ship's performance. The manual graphic system, described in the previous section, does not cater for the speed and consumption relationship, and a further graphic system is necessary as shown in Figure 14.

As will be seen, it is self-explanatory and is usually on the basis of fair weather. Actual service results can be shown and the author favours a recording method with all weather and fair weather corrected figures. Deviation from the curves can be indicated either vertically (as fuel consumption) or horizontally (as speed) difference. They can be updated as and when differences in performance are such that the existing curves give unattainable performance. For example, if the vessel's bottom is fouled.

Sea trials performance, converted from horsepower to fuel consumption at the expected fuel quality, can be shown for reference and the sea margin easily calculated for predictive purposes.

For reference, a graphic illustration of four typical vessels giving fair weather service performance of modern designs compared with older designs showing the quite remarkable improvements that have been made in the intervening period is provided (see Figure 15).

Fig. 14.   Speed and fuel consumption graph — Panamax

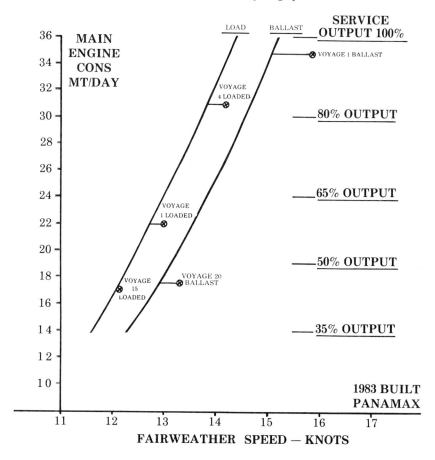

10. Summary of Chapter Three

There are many expensive performance monitoring systems on the market and they no doubt do a good job. For those not wishing to spend any money on additional equipment, the simple method described here may suit their purpose. In a later chapter, a low cost computerised performance system will be described.

Fig. 15.   Speed and fuel consumption graph — various vessels

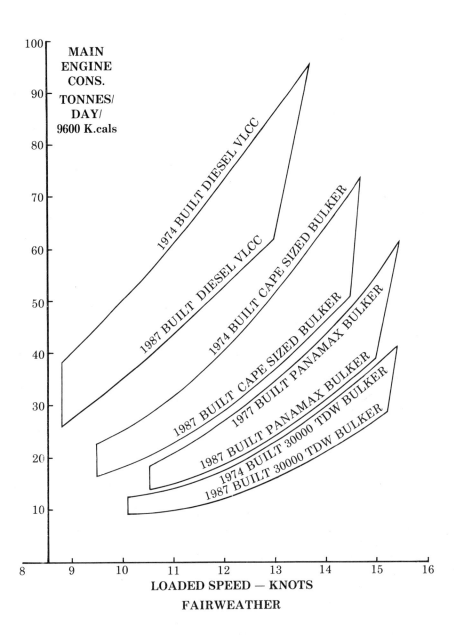

LOADED SPEED — KNOTS

FAIRWEATHER

CHAPTER FOUR

Factors affecting choice of a vessel, its machinery and equipment

1. Basic philosophy

Shipowners intending to order new tonnage must consider every aspect in what should be a controlled decision-making process. In the real world this seldom happens and new vessels are, in many cases, ordered for seemingly obscure reasons. A lot depends on whether the proposed newbuilding is for a long-term time charter, arranged coincident with signing a newbuilding contract, for speculative time charters or for operating on the spot market or even for an affreightment contract. It could even be for bareboat charter purposes.

Each possibility will be discussed on its merits, but it is assumed that no prior agreement with any shipyard has been made. It is also assumed that the shipowner has already made a choice of the basic type of vessel he requires. Other aspects will also be considered which affect the vessel's operational performance.

2. Long-term time charter

This type of agreement used to be popular especially with Japanese trading houses, who would arrange both the shipyard and charterer. In recent years these agreements have fallen from favour and are unlikely to reappear until freight rates are restored to profitable levels.

The charterer will usually make all the major decisions regarding cargo gear, speed, type of cargoes, cubic capacity, range and geographical area of operation. The shipowner will make decisions relating to Main Engine and auxiliary machinery selection, flag state, classification society and various other items. For the purpose of our study, we will assume that the choice of shipyard has been left to the shipowner without any interference from the charterer.

The technical department of the shipowner should, ideally, be aware of all shipyards capable of building the proposed vessel and a telexed enquiry giving outline particulars, delivery requirements and asking for a price indication will be sent to those shipyards. About 30 shipyards are usually circulated and, in the current climate, most will reply giving particulars of their standard design closest to the shipowner's outline. From these replies a short list of three will probably be drawn up, based on previous experience with the shipyard or at least in the country where it is located. This will normally give an indication of the financial terms available and the workmanship expected.

At this stage, the shipowner's special requirements will be introduced to the shortlisted shipbuilders. High on the list will be the performance of the vessel, both in technical and commercial terms. On the technical side propulsion efficiency will receive priority also auxiliary energy consumers, especially if a product carrier or a crude tanker is involved. As an example of this we could examine the operating scenario of a typical product carrier. If the shipowner's/charterer's chartering department is efficient, a very low percentage of the vessel's sea time will be spent in the ballast condition. Emphasis must, therefore, be given to efficient tank cleaning systems and cargo tank heating requirements; both high energy consumers. Cargo discharge operations must also figure high on the list for this type of vessel to reduce inport time.

A modern product carrier will almost certainly have an electro-hydraulic centralised system supplying the needs of both cargo pumps and deck mooring machinery. The improvement in performance using this type of auxiliary equipment over the previous method of using a conventional pumproom housing steam-driven cargo pumps and with steam mooring winches is quite dramatic. A cargo discharge operation on a 30,000 TDW product carrier, using the conventional system, consumes around 50 tonnes heavy oil; a modern electro-hydraulic system, perhaps 12 tonnes of blended fuel. Also, energy for the system is immediately available at the touch of a button instead of waiting until a boiler is flashed; itself an energy waster.

In the case of Panamax bulkers, there is no possibility of such ancillary system saving. Even so, centralised electro-hydraulic deck machinery systems have been commonplace for many years on Panamax bulkers and it was a logical development to extend their use to tankers.

Returning to our product carrier, cargo heating requirements can be rather high and are not normally subjected to charterparty performance warranties. However, it is in owner's and charterer's interest to minimise fuel consumption in this critical area. One means of achieving this is to provide the largest possible exhaust gas economiser so that surplus heat in the exhaust gases can be converted to steam and directed to cargo heating purposes at sea. Any extra cost for the provision of this additional plant would be the subject of a financial agreement between shipowner and charterer.

Regarding charterparty speed and consumption warranties, the figures agreed between shipowner and charterer will determine the horsepower requirements. In the case of a bulk carrier using a standard NYPE form, it is usually specified that the warranted performance is in fairweather conditions. Shipowners, therefore, have an opportunity of using a lower sea margin to calculate the vessel's horsepower requirements. By using a sophisticated self-polishing underwater paint system, the hull-fouling contribution to the sea margin can theoretically be somewhat reduced.

We could say that if a NYPE charterparty is involved the horsepower requirements could be downgraded by a contribution of the fairweather clause and the use of self-polishing paint. It is becoming increasingly popular to define weather conditions in an additional clause forming part of the charterparty as a further protection to owners against performance claims.

So instead of using, say, a 20% sea margin for an unqualified performance, the shipowner could perhaps opt for a 10% sea margin. This represents a very considerable saving in the vessel's construction costs reflected by having a much smaller Main Engine. A smaller Main Engine means less space required for the engine room, as well as smaller pumps, generators, even smaller pipes, electric cables and a host of other items — not forgetting the benefit of increased cargo deadweight.

Should the proposed vessel not have the benefit of a fairweather clause in the charterparty, the advantages mentioned above would not, of course, apply. The shipowner, based on his own experience, would have to decide what percentage sea margin should be included and most shipbuilders will produce simple graphs showing speed and consumption with various sea margins as shown in Figure 16.

Fig. 16.   Speed and fuel consumption graph showing sea margin

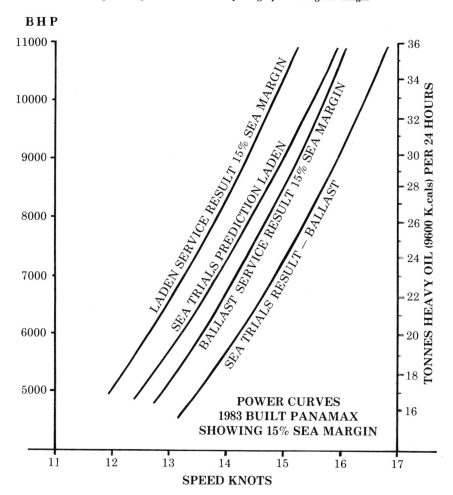

3. Speculative charters and spot voyages

In the case of this type of operation, the shipowner will not be able to rely on the protection of a fairweather clause unless standard NYPE charterparty terms are envisaged throughout the vessel's economic life. The shipowner does, however, have a good opportunity to decide what size Main Engine he should specify. We have seen the advantages of having a smaller Main Engine, and an actual example of this is now given.

If we take a typical Panamax bulker — ordered in 1974 and delivered in 1977 — it was then the vogue for this type of vessel to be rather highly powered. In the example under review the installed horsepower was provided by a 17,000 BHP Main Engine which was not uncommon. The vessels (two were ordered) entered service and immediately commenced slow steaming, as their entry into service coincided with a trough in the Panamax market.

They slow steamed at around 6,000 BHP until 1981 and throughout that year operated at 65% power — equal to around 11,000 BHP. During 1982 they reverted to slow steaming, and have operated this way until the present time (fourth quarter 1986).

In 1981, when the Panamax market looked promising, four additional Panamax bulkers were ordered. The first two were, more or less, standard designs and were provided with 14,000 BHP Main Engines. It was not possible to reduce the size of the engine nor was it possible to obtain the latest derated engine. The opportunity presented itself to take this course of action with the later vessels, and 11,000 BHP derated long stroke engines were provided. This, more or less, tied up with the horsepower requirements during the good year of 1981.

The arrival in service of these four Panamaxes also coincided with another trough in the market (1982–83). All four vessels have generally been slow steaming, except for isolated voyages on the most recently delivered vessel which is time chartered. So, the 1977-delivered Panamaxes have never used their 17,000 BHP. The 14,000 BHP Panamaxes have never used their service power either, and the 11,000 BHP Panamaxes only on one or two voyages. It should be mentioned that slow steaming on the later two Panamaxes is only around 4,000 BHP.

Not many Panamaxes have been ordered of late, but the few that have been still appear to be overpowered, having Main Engines of around 9,000 BHP. At this sort of installed horsepower, laden speeds of 14.5 knots, with a consumption of 30 tonnes per day, would be achievable. The question to be answered is will this speed be more than sufficient for the requirements of the next decade, or can we afford to further reduce the installed horsepower even more?

One important item occasioned by the reduction in Main Engine output was the elimination of the forward bunker tank and its sometimes troublesome heating and transfer system.

4. Charter of affreightment

In a very simplistic charter of affreightment contract, a shipowner will ship an agreed quantity of cargo from A to B on an annual basis over a period of time. If the period is sufficiently long a newbuilding may be the most cost-beneficial solution to the problem, especially if existing vessels are not ideally suitable. The shipowner has a good opportunity to build the most suitable economic vessel if this course of action is justifiable, given the length of contract.

The project can be approached using a selection of vessels with varying cargo deadweights and installed power plants. The loading and discharge port physical restrictions will probably be a limiting factor, unless some interport restriction, i.e. canal transits, overrides the port restrictions. There may be a necessity to use a port state registry, which could greatly affect operating costs, and this aspect must be carefully examined.

Shipbuilding costs, like freight rates, follow some unpredictable cyclic pattern — as do fuel costs. It is, therefore, very difficult to arrange things so that all financial aspects fall into place to the benefit of the owner or charterer. An exercise undertaken some time ago at the ruling shipbuilding, operating and fuel costs is given in Table 8 and updated figures are shown for comparison. These updated figures do not fully take into consideration the improved performance now available, but simply serve as a comparison.

Table 8.   Charter of affreightment calculations

	1982 study		Updated to 1986	
	Vessel A	Vessel B	Vessel A	Vessel B
Cargo deadweight	37,500	33,500	37,500	33,500
Service BHP	7,600	12,800	7,600	12,800
Service speed	13.75	15.75	13.75	15.75
Vessel cost	$22.4M	$23.2M	$14M	$14.7M
Fuel cost per tonne $	175/300	175/300	70/125	70/125
Daily fuel cost — 365 days $	4,230	6,580	1,692	2,632
Daily finance cost $	13,360	13,801	8,350	8,745
Daily operating cost $	3,760	4,000	2,460	2,700
Total daily cost $	21,350	24,381	12,502	14,077
Required freight rate	$26	$29.70	$15.20	$17.12

The basic requirement of this particular charter of affreightment was that the annual quantity shipped was 300,000 tonnes over a total distance per voyage of 12,660 miles. Six different combinations of deadweight and horsepower were included in the study; only the best and worst are shown in Table 8. The study illustrates that the cost of speed is rather high and the largest deadweight, smallest powered vessel was the most cost-effective — even though large differences in fuel costs are taken into consideration. Each shipowner will have his own ideas for arranging finance; those shown may have to be adjusted on whatever terms are currently available in the country where the shipyard is located.

5. Bareboat charter

It is rather difficult to formulate any positive policy with regard to bareboat charters. Whoever puts up the finance gets his return on investment and is not directly concerned with performance during the period of the charter. The ship operator is concerned with the performance of the vessel but does not control the specification of the vessel to the extent he would if it were his own.

Bareboat charters are generally arranged with fixed business in mind and, providing the ship performs within the warranted speed and consumption, no problems should arise.

Any additional features required by the operators would have to be passed on to them, presumably in the form of increased hire.

6. Cargo related operations

To enhance the performance of any vessel it is important to minimise the amount of off-hire time when performing essential cargo space cleaning operations. Dealing first with bulk carriers employed in the grain trade, it is quite common for the hold spaces to be turned down by a surveyor from a cleanliness point of view. Some of these decisions are questionable, but the shipowner's stance should be to make sure there is no room for argument by spending sufficient time and effort on the cleaning operation. In certain parts of the world, for example Australia, it is not permitted for crew members to clean holds — in the event of these being turned down by a port official. Shore labour has to be used and is very expensive in that part of the world.

Reduced crew numbers do no help matters, and every assistance should be given them in the execution of this task. Portable, even permanent, hold cleaning machines using water and compressed air are available on the market. To assist in removing the washed-down residues permanent piping to the upper deck can be installed so that a mucking-out pump is quickly connected without having to haul long lengths of hose around. Small portable davits for each hold access opening to asssist in handling the mucking out pumps and drums used for final cleaning operation are more or less essential.

For bulkers carrying grain in topside tanks it is a good idea to arrange the longitudinal stiffeners in the hold rather than in the tank with the standing flange positioned downwards sloping in such a way as to avoid trapping any grain. This action will greatly speed up cleaning operations and will also help avoid build-up of mud in the topside tanks which usually leads to the start of serious corrosive action. If carrying Petcoke, the introduction of a chemical into the washing water has proved successful.

Bilge wells in cargo spaces can be arranged with a duplex well having a weir interposed to reduce cleaning out operations and also help to keep residues out of the bilge piping system. The capacity of the bilge pumps must meet classification rules, but should also be capable of dealing with large quantities of hold cleaning water.

To improve the flexibility of bulk carriers, consideration should be given to the provision of special equipment for carrying cargoes of a dangerous nature. Direct reduced iron (DRI) comes into this category, and from time to time enquiries are received for vessels fitted with special nitrogen or CO_2 blanketing arrangements, enabling this product to be carried.

Various seed cakes and pellets also require to be protected by either a fixed CO_2 total flooding system or, in some instances, a portable high expansion foam generator — depending on the physical characteristics of the cargo. The carriage of sulphur requires special precautions, and many shipowners exclude this cargo when drawing up a charterparty. Contact with water, especially sea water, is to be avoided and hold spaces should ideally be protected by paint and then limewashed before loading.

In the case of crude oil carriers, MARPOL Annex I regulations cover the cleaning of cargo tanks, and little can be added. The advent of crude oil washing has certainly had an impact on reducing the amount of "digging out" found necessary prior to the introduction of this MARPOL regulation. Segregated ballast tanks have also played a very important part in reducing tank cleaning operations.

Product carriers and similar types of vessel generally have different cleaning requirements in that commercial considerations outweigh MARPOL regulations. The standard of cleanliness to meet charterers' inter-cargo requirements is high, and efficient tank cleaning equipment must be of the highest calibre. In the case of edible oil and similar viscous cargoes, the labour-intensive operation of "sweeping out" can be lessened by the introduction of inboard tank surfaces being completely free from all stiffeners and heating coils. This is usually arranged by having a double bottom tank in which these stiffeners are located. Recently delivered vessels of this type have the upper deck stiffeners mounted actually on the upper deck instead of in the tank.

Heating coils can be eliminated by the provision of independent deck mounted heaters through which the cargo is circulated by means of the cargo pumps. This greatly reduces tank cleaning operations, but other aspects should be considered; for example, running the cargo pumps at sea. An example of a modern product carrier is shown in Figure 17.

Accessibility to the upper parts of tank structures has always been a problem for inspection purposes relating to both cargo and classification surveyors' requirements. There are "knock down" lightweight staging frames available which are capable of being

Fig. 17.   Tank arrangement — modern product carrier

THE DIFFERENCE

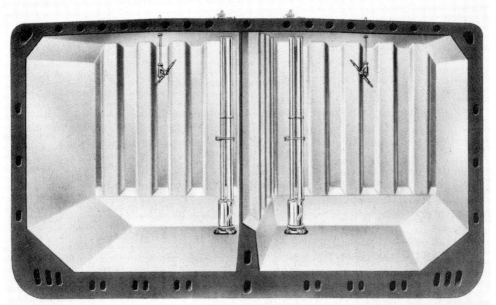

YOUR CHOICE

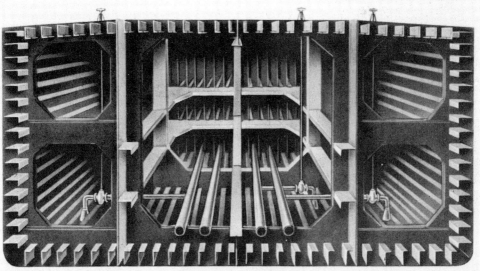

ALTERNATIVE???

passed through the tank entry opening and assembled *in situ*. Another possibility is to use a special raft which can be inflated on the tank bottom when empty and floated up when ballasting the tank to whatever level requires inspection.

Chemical carriers have their own special cleaning requirements and pollution aspects are covered by the MARPOL Annex II regulations which entered force in April 1987. Tank structures have an even greater possibility of being arranged without internal stiffeners on these specialised vessels. On more sophisticated chemical carriers stainless steel is used extensively, as are deck mounted tanks.

7. Labour saving devices

With the current trend to reducing the numbers of ship's staff for survival, let alone economic reasons, it is very important that the remaining staff can carry out labour-intensive operations efficiently. Mooring equipment is a case in point and, arguably, was the last obstacle to making any significant reduction in the deck manning level. Self-tensioning winches and remote centralised winch controls were introduced, as were large barrels which enabled mooring ropes to be permanently coiled on the winches rather than in underdeck lockers. Another feature gaining popularity is the bow thruster which not only simplifies mooring operations but also makes a significant reduction in the number of tugs required each time the vessel enters or leaves port.

Easy access to store spaces using gantry type or radial cranes, instead of a rigged derrick, also saves many man hours. Some vessels on dedicated runs were able to use the container method of storing with some success. In this scheme, a full container is loaded at the home port and any remains in the landed container credited to the shipowner.

Access to the engine room and also within the machinery spaces to permit the easy handling of heavy parts is rather important nowadays. Tremendous advances have been made in the development of hydraulic and pneumatic tools permitting one-man operation for dismantling procedures which previously required a gang of men.

The use of superior materials in many machinery components leads to longer operational life and much less maintenance. This would also apply to the use of superior lubricating oils which extend operational intervals between piston overhauls. Now that installed Main Engine powers have dropped, it is possible to specify a fewer number of cylinders — the author was instrumental in using four cylinder large bore engines, which reduces maintenance considerably.

The introduction of self-cleaning heavy oil purifiers also made the labour-intensive function of dismantling and cleaning purifiers by hand a thing of the past. In these days of poor quality fuel, self-cleaning purifiers have come into their own, not only from a labour-saving point of view but also from consideration of the problem of hygiene.

Unmanned engine rooms are now standard and the main benefits are to release engine room personnel from tedious watchkeeping duties and redeploy them in carrying out essential maintenance functions as well as allowing reductions in complement. It is important to see that engine room workshops are properly equipped and ship's personnel adequately trained in workshop and welding techniques so as to be able to take advantage of the increased amount of time available for maintenance functions.

Planned maintenance, if carried too far, can be counter-productive in that increased spare parts consumption can rise dramatically to meet the sometimes excessive demands of the maintenance schedules. When freight rates are high off hire must be avoided, when they are low it is, of course, not so important. Planned maintenance became popular in the heady days of high freight rates and its significance as part of the overall strategy is somewhat reduced, and will continue to be so until freight rates recover.

Maintenance of steelwork is an area which should receive care, preferably at the vessel's construction stage. Paint coatings such as zinc silicate can virtually reduce fabric maintenance of the upper deck steelwork to zero. Members of the deck crew can then be employed in maintaining the cargo spaces and associated equipment which are revenue related, as opposed to fabric maintenance usually a cosmetic, time-consuming chore.

Tank coatings should also receive careful consideration in relation to their required duty. This would apply to cargo tanks, ballast tanks and cargo holds. Cathodic protection in ballast tanks is an added insurance against corrosion. This develops rapidly in coated tanks when small coating breakdowns direct all the available potential onto the unprotected area, which experiences rapid deterioration in the form of deep pits. The extensive use of plastic-type hardwearing accommodation bulkheads avoids the repainting of these spaces over a normal vessel's lifespan and, therefore, makes a useful contribution to labour-saving.

8. Registry and manning considerations

The commercial performance of any vessel is closely related to the country of registry and nationality of the officers and crew. We have learnt, in previous chapters, of the tremendous improvements in overall vessel performance expressed in miles per tonne of fuel used, and there is little scope for any further dramatic improvements. This is not the case with respect to registry and manning, and the attention of shipping economists has been transferred from the technical scenario to these important aspects.

Historically, vessels were registered in the port in which the owner was located and crewed by personnel residing in or close by. The emergence of the Panamanian and Liberian registers changed all that and there are very few vessels left which operate in the traditional manner. There are, however, still some countries which apply protectionist policies to their shipping industries, but these are not within the scope of our study.

It is, therefore, assumed that the shipowner has a fairly flexible approach to registration and manning and is not inhibited by too many constraints. The first decision is, probably, whether to manage the vessel himself or to enter into a management contract with one of the large number of specialised companies now existing and growing at an increasing rate. These management companies will give various management options such as crew only, officers and crew, full operational and, finally, operational and chartering management.

The shipowner may have his own staff capable of carrying out all these functions himself and may choose this option. Management companies do have access to low-cost officers and crew and a shipowner may not have such access. They also have the possibility of arranging bulk discounts on all manner of purchases; again the shipowner may not have this option, purely on account of the reduced size of his fleet. All management companies give their shipowner customers regular computerised accounts, showing actuals against budget and accrual rate for all the various operating expenses.

In order to remain competitive with a management company there are various measures a shipowner can take. For example, he can encourage his officers to sign an "offshore" agreement whereby the officer is responsible for most, if not all, of those additions to his salary normally borne by the shipowner. The main benefit in this scheme is the reduced amount of leave taken, with a corresponding reduction in repatriation costs and coverage of the rank. In extreme cases this has risen to two personnel per rank but can be stabilised at much less than this under the "offshore" agreement.

There are many crewing agencies who will supply crew members only and, providing they speak the same language as the officers, no problems normally arise.

It is unlikely that European owners can reduce manning costs to those available using a management company employing Far Eastern officers and crew. Some charterers, however, have objections to using this type of crewing operation and this could be a stumbling block to their wholesale use.

For shipowners having a rather large fleet it is possible to give one or two vessels to a management company on a trial basis, so that results can be monitored and compared with existing sister ships. This has an added attraction in that the original staff will produce their best endeavours to show they are capable of matching the opposition.

Choice of registry also has an input to the equation. The main benefits in registering in a so-called flag of convenience country are well known and no further comment is called for.

Normally the port of registry will not affect the cost of a newbuilding as most shipbuilding nations are signatories to the IMO Conventions governing safety, pollution and other such matters. This is not the case with respect to manning, especially the actual numbers of officers/crew to be accommodated aboard. Not so long ago, a total complement of 40 would be aboard our standard vessel. This has gradually been eroded and now probably averages around 30 on a world-wide basis and very considerably less in specially designed newbuildings. The size of the accommodation deckhouse can be considerably reduced to meet the needs of the smaller number of crew members, with a resultant saving in building costs.

9. Secondhand tonnage

Purchase of secondhand tonnage is rather restrictive with regard to performance aspects, although previous chapters will give an idea of how to judge expected performance based on examination of the vessel's logs. The results can then be compared with data given earlier for the various types of vessels, particularly the fuel coefficient. It is possible to provide some of the performance improvers, should this course of action be deemed desirable after studying the previous performance results.

One of the most important aspects when contemplating the purchase of a secondhand vessel is to spend some time perusing the logs, abstracts and work books. If these have been filled in correctly, a wealth of information relating to the vessel's performance can be extracted. This would also apply to the cargo record book which can give valuable information for the use not only of the buyer's sea staff but also the chartering department.

Classification records will give details of the various surveys carried out both for regulatory and damage purposes which will give a good idea of the vessel's condition and expected performance reliability in service.

A very useful service is provided by Lloyd's Intelligence at Colchester, UK, who, for a nominal fee, will give all reported incidents to any vessel under consideration for purchase.

10. Commercial aspects

Given the cyclic nature of the shipping industry, it is very important that the timing of any shipbuilding contract is judged correctly. When the market is buoyant shipyard order books are full and the time-span between signing a contract and taking delivery of a new vessel can be three years. This is an almost impossible commercial restriction for any shipowner to be faced with when attempting to arrange a time charter so far in advance.

The question of escalation also rears its ugly head and, in the situation described, shipbuilders either close their order books or insist on an escalation clause, usually related to steel and manpower costs.

In a depressed market the ball game changes completely and the time span between signing contract and accepting delivery can be as short as one year. It is at these times that the speculators move in to sign contracts for ships at about half the cost of that ruling in the buoyant period. Panamax bulkers ordered in 1981 and delivered in 1983 cost around $30m each and probably required a daily charter rate of $16,000 to give a reasonable return on this amount of investment. This type of vessel, if ordered in 1984, could have been obtained for around $18m. Several Panamaxes delivered in 1983 were, in fact, sold by the banks, who had repossessed them, for around $9m "as is where is" in October 1985.

Such are the vagaries of the sale and purchase market which is obviously not for the faint-hearted. When the time between contract and delivery is as short as one year it follows that, with the enormous shipyard capacity available, any but the most unforeseen recovery will be short-lived.

11. Summary of Chapter Four

The approaches which can be made to the various types of charterparties are illustrated. Shipowners, in order to remain competitive, must provide the latest labour-saving devices to assist in reducing crew members. Manning costs are the current target for the economists, and management companies are the current fashion for reducing these costs. Owners with large fleets can partially compete by arranging offshore agreements and employing low-cost ratings.

The timing of a shipbuilding contract is important; on balance vessels ordered at peaks in the market would appear to have the tremendous financial disadvantage of never repaying this cost over the vessel's life cycle.

CHAPTER FIVE

Charterparty performance claims

When the freight market is depressed charterparty performance claims abound, especially when the vessels concerned have been fixed at a high rate and the charterers try to retrieve the situation. It can become a little complicated when vessels are sub-chartered and the sub-charterer claims from the head charterer who then claims from the shipowner. Each basic type of charter, bulker and tanker has different approaches to the question of performance claims and these will be reviewed.

1. Bulk carriers

A large number of bulk carriers are chartered under the New York Produce Exchange form (NYPE) which has been in use for many years. Latterly, additional clauses have been added to cover performance at various fuel consumptions lower than the single fuel consumption hitherto used.

In the standard NYPE charterparty form, the relevant clause relating to performance is: "... capable of steaming, fully laden, under good weather conditions about ... knots on a consumption of about ... tons of best Welsh Coal ... best grade fuel oil ... best grade Diesel Oil ..."

It will be seen that the legal interpretation of the word "about" would figure highly in any claim for under-performance. Regarding the vessel's speed, it would appear that "about" has been arbitrated as being within half a knot tolerance. With respect to fuel consumption, it is not widely published what the tolerance is but 3% seems to be a figure frequently used, although cases of 5% have been known.

Turning to the interpretation of fairweather, unless this is spelt out in an additional clause, days when wind forces of Beaufort 5 and above are recorded are normally allowed to be excluded from the calculation submitted for an under-performance claim. Here, again, cases in which Beaufort force 4 and above was allowed have been heard of.

So how are these underperformance claims calculated? If we take a straightforward example to start with, a vessel on a voyage from New Orleans to Rotterdam averaged 11.62 knots on 28 tonnes of heavy fuel oil and 2.2 tonnes of diesel oil per day. The NYPE charterparty speed and consumption was 12.5 knots on 27 tons of heavy fuel oil and two tons of diesel oil. It should be noted that the vessel used metric units of measurement and the NYPE form used imperial units. It is not unknown for American units (short tons) to be used and whilst the differences are small, they must be taken into consideration.

We can start by converting the charterparty figures to metric as under:

27 tons = 27 × 1.01605 = 27.43 tonnes

2 tons = 2 × 1.01605 = 2.03 tonnes

Passage distance = 5,000 nautical miles

Charterparty speed = 12.5 knots − 0.5 allowance = 12 knots

$$\text{Charterparty time} = \frac{5{,}000}{12} = 416.67 \text{ hours}$$

$$\text{Actual time} = \frac{5{,}000}{11.62} = 430.29 \text{ hours}$$

Loss of time = 13.62 hours

Hire rate = $2.60

Deadweight = 60,460 tons

$$\text{Hourly hire rate} = \frac{60{,}460 \times 2.6}{30.4 \times 24} = \$215.45$$

In this case, the average of 30.4 days per month is used, which appears to be standard for time charters of six months or more. In the example under review, the under-performance claim for increased passage time would therefore be:

(1) 13.62 × 215.45 = $2,934.43

We must also consider the additional fuel consumed both by virtue of the increased passage time and, in this particular case, the increased daily consumption over that warranted in the charterparty. This is calculated as follows :

$$\text{Charterparty H.O. consumption (with 0.5 knot and 3\% fuel)} = \frac{5{,}000}{12 \times 24} \times 27.43 \times 1.03 = 490.50 \text{ tonnes}$$

$$\text{Actual consumption} = \frac{430.29}{24} \times 28 = 502 \text{ tonnes}$$

(2) Additional fuel consumption = 11.5 tonnes

The same applies for diesel oil and in this particular example would be:

$$\text{Charterparty D.O. consumption (with 0.5 knot and 3\% diesel)} = \frac{5{,}000}{12 \times 24} \times 2.03 \times 1.03 = 36.3 \text{ tonnes}$$

$$\text{Actual diesel consumption} = \frac{430.29}{24} \times 2.2 = 39.44 \text{ tonnes}$$

(3) Additional diesel consumption = 3.14 tonnes

One further point we should take into account is a refund on brokerage relating to the time element only. We are now in a position to calculate this claim and we will assume the brokerage was 3.75% of the cost of heavy oil at $70 and of diesel oil at $170 per tonne. The calculation will then become:

		$
(1)	Excess hire	2,934.43
	3.75% address commission	(110.04)
(2)	Excess heavy oil 11.5 × $70	805.00
(3)	Excess diesel oil 3.14 × $170	533.80
	Total	$4,163.19 under-performed

This example describes what might be called a typical case of a Panamax bulker bringing a cargo of grain from New Orleans to Rotterdam. It would be extremely unlikely that a passage of this nature would be performed in good weather conditions as described in the NYPE charterparty. We must, therefore, adjust the above claim by including some heavy weather and, in order to explain the procedure fully, an abstract covering the voyage in question is shown in Table 9.

Ship Performance

Table 9. Passage summary — all weather

Date	Distance	Time	Speed	Wind force	RPM	H.O.	D.O.	Remarks
1.10.86	292	24.00	12.17	2	93.2	28.0	2.44	
2.10.86	287	24.00	11.96	3	93.1	28.0	2.2	
3.10.86	299	24.00	12.46	3	94.1	28.0	2.2	
4.10.86	237	24.00	9.87	7	90.6	27.5	2.2	
5.10.86	238	24.00	9.92	7	90.5	27.5	2.2	
6.10.86	242	24.00	10.08	5	91.2	27.5	2.2	
7.10.86	252	24.00	10.50	5	93.1	28.0	2.2	
8.10.86	293	23.00	12.74	3	93.8	27.0	2.2	Clocks advanced
9.10.86	285	23.00	12.39	2	92.7	27.0	2.0	Clocks advanced
10.10.86	298	23.00	12.96	1	93.4	27.0	2.0	Clocks advanced
11.10.86	282	23.00	12.26	4	93.1	27.0	2.0	Clocks advanced
12.10.86	294	23.00	12.78	4	93.6	27.0	2.0	Clocks advanced
13.10.86	271	23.00	11.78	5	92.4	26.5	2.0	Clocks advanced
14.10.86	259	24.00	10.79	6	91.8	27.5	2.2	
15.10.86	301	24.00	12.54	4	92.7	28.5	2.2	
16.10.86	253	24.00	10.54	6	91.3	28.0	2.2	
17.10.86	272	24.00	11.33	5	91.7	28.0	2.2	
18.10.86	292	24.00	12.17	4	93.1	28.5	2.2	
19.10.86	53	4.18	12.32	4	93.4	5.5	.6	
Totals	5,000	430.18				502	39.44	
Averages			11.62		92.8	28.0	2.20	

This is the complete voyage showing every day's performance, and we can show the fairweather figures by eliminating those days during which the average wind force exceed Beaufort 4 as follows:

Table 10. Passage summary — fairweather

Date	Distance	Time	Speed	Wind force	RPM	H.O.	D.O.	Remarks
1.10.86	292	24.00	12.17	2	93.2	28.0	2.44	
2.10.86	287	24.00	11.96	3	93.1	28.0	2.2	
3.10.86	299	24.00	12.46	3	94.1	28.0	2.2	
8.10.86	293	23.00	12.74	3	93.8	27.0	2.2	
9.10.86	285	23.00	12.39	2	92.7	27.0	2.0	
10.10.86	298	23.00	12.96	1	93.4	27.0	2.0	
11.10.86	282	23.00	12.26	4	93.1	27.0	2.0	
12.10.86	294	23.00	12.78	4	93.6	27.0	2.0	
15.10.86	301	24.00	12.54	4	92.7	28.5	2.2	
18.10.86	292	24.00	12.17	4	93.1	28.5	2.2	
19.10.86	53	4.18	12.32	4	93.4	5.5	.6	
Totals	2,976	239.18			93.4	281.5	21.04	
Averages			12.44			28.23	2.11	

Returning now to the all weather calculation, we can substitute the fairweather averages as indicated above.

Actually, there is another factor to be taken into account based on arbitrations, which effectively means that if a vessel is underperforming in fairweather it would similarly underperform in heavy weather. We must, therefore, apply the average fairweather performance to the whole voyage. As it happens, the example we have chosen does not clearly indicate this point, as the vessel was overperforming in fairweather. However, we will calculate the performance claim just to illustrate how it is done:

$$\text{Fairweather time} = \frac{5,000}{12.44} = 401.93 \text{ hours}$$

$$\text{Charterparty time} = \frac{5,000}{12} = 416.67 \text{ hours (as original)}$$

Using the ½ knot tolerance, the vessel clearly overperformed and no claim would arise with respect to time lost.

The H.O. fuel consumption will also be calculated:

$$\text{Fairweather consumption} = \frac{401.93}{24} \times 28.23 = 472.77 \text{ tonnes}$$

Charterparty consumption = 490.50 tonnes (as original)

The vessel clearly underconsumed H.O. and no claim would be forthcoming from charterers. In the event of the loss of time being considerable it can be offset by savings in fuel consumption, but this is clearly not the case here.

D.O. will likewise be calculated:

$$\text{Fairweather consumption} = \frac{401.93}{24} \times 2.11 = 35.33 \text{ tonnes}$$

Charterparty consumption = 36.3 tonnes

Again, the vessel underconsumed and no claim would be expected.

It will be seen that by applying the good weather clause, an underperformance claim of $4,163.19 can be reduced to nothing.

As a point of interest, we can calculate the amount of underperformance if no good weather clause was used nor were the ½ knot and 3% tolerances applied:

$$\text{Charterparty time} = \frac{5,000}{12.5} = 400 \text{ hours}$$

Actual time = 430.29 hours

(1) Loss of time = 30.29 hours

$$\text{Charterparty H.O. consumption} = \frac{5,000}{12.5 \times 24} \times 27.43 = 457.17 \text{ tonnes}$$

Actual H.O. consumption = 502.00 tonnes

(2) Additional H.O. consumption = 44.83 tonnes

$$\text{Charterparty D.O. consumption} = \frac{5,000}{12.5 \times 24} \times 2.03 = 33.83 \text{ tonnes}$$

Actual D.O. consumption = 39.44 tonnes

(3) Additional D.O. consumption = 5.61 tonnes

The calculation then becomes:

(a)	Excess hire 30.29 × 215.45	=	$6,525.98
	3.75% address commission	=	($244.72)
(b)	Excess heavy oil 44.83 × $70	=	$3,138.10
(c)	Excess diesel oil 5.61 × $170	=	$953.70
	Total		$10,373.06

We can now examine what all this means in terms of charterparty clauses and its effect on the shipowner using the voyage described earlier as the basis in Table 11.

Table 11.   Summary of charterparty clauses

Summary of clauses	Fuel costs $	Daily hire $	Under-performance claim $	Cost/day $
All weather, no speed or speed allowances	70/170	5,170	10,373.06	578
All weather, half knot and 3% fuel allowance	70/170	5,170	4,163.19	232
Good weather, half knot and 3% fuel allowance	70/170	5,170	Nil	Nil

Should fuel costs have been at their 1985 level the situation would, of course, have been much worse. If, as sometimes happens, shipowners are obliged to accept an all weather clause and/or guaranteed speeds and consumptions it will serve as a reminder of what sort of claims they are liable for. The highest underperformance claim under an all weather clause seen by the author in recent years amounted to $67,152 on a single voyage — Australia to Europe. Should the shipowner have had the benefit of a good weather clause there would have been no underperformance claim, as the vessel was within the warranted figures. Instead, a claim amounting to $1,160 per day had to be settled by the owner.

On occasions, a voyage may be performed at two distinct speed settings at charterers' instructions, and the calculations should be on the basis of two separate voyages, not on the average of the whole. Should more than two speed settings be used, it may be in the shipowner's favour to discount this voyage.

On other occasions vessels are asked by charterers to perform the voyage at a fuel consumption figure between those specified in the charterparty. The underperformance claim is then submitted at a fuel consumption specified in the charterparty. This is a potentially dangerous situation in financial terms, and an example of how the unsuspecting shipowner can incur a penalty is given.

Charterparty warranted performance:

> 13 knots on 48 H.O. $\times$ 2.0 D.O.
> 12 knots on 38 H.O. $\times$ 2.0 D.O.
> Actual fairweather average: 11.44 knots on 43.3 H.O. $\times$ 1.2 D.O.

As will be seen, this is between two consumption settings.

The voyage in question was 5,597 miles and the hire rate was $288 per hour and fuel costs $182/$250. Charterers chose the 38 T/D performance to calculate the claim.

Charterers' calculations

$$\text{C/P time} = \frac{5,597}{11.5} \times 486.7 \text{ hours}$$

$$\text{Actual time} = \frac{5,597}{11.44} \times 489.2 \text{ hours}$$

$$(1) \quad \text{Excess} = 2.5 \text{ hours}$$

$$\text{C/P H.O.} = \frac{486.7}{24} \times 38 \times 1.03 = 793.8 \text{ tonnes}$$

$$\text{Actual H.O.} = \frac{489.2}{24} \times 43.3 = 882.6 \text{ tonnes}$$

$$(2) \quad \text{Excess} = 88.9 \text{ tonnes}$$

$$\text{C/P D.O.} = \frac{486.7}{24} \times 2.0 = 40.6 \text{ tonnes}$$

$$\text{Actual D.O.} = \frac{489.2}{24} \times 1.2 = 24.5 \text{ tonnes}$$

(3) D.O. savings = 16.1 tonnes

(a)	2.5 × $288	=	$720
	Commission	=	($27)
(b)	88.9 × $182	=	$16,180
(c)	16.1 × $250	=	($4,025)
	Net underperformance	=	$12,848

Owners' calculations

Fairweather average 11.44 on 43.3/1.2, corrected for 38 T/D using cube law:

$$\sqrt[3]{\frac{11.44^3 \times 38}{43.3}} = 10.95 \text{ knots}$$

C/P time = 486.7 hours

$$\text{Actual time} = \frac{5,597}{10.95} = 511.1 \text{ hours}$$

(1) Excess = 24.4 hours

C/P H.O. = 793.7 tonnes

$$\text{Actual H.O.} = \frac{511.1}{24} \times 38 \times 1.03 = 833.5 \text{ tonnes}$$

(2) Excess = 39.8 tonnes

C/P D.O. = 40.6 tonnes

$$\text{Actual D.O.} = \frac{511.1}{24} \times 1.2 = 25.5 \text{ tonnes}$$

(3) D.O. savings 15.1 tonnes

(a)	24.4 × $288	=	$7,027
	Commission	=	($261)
(b)	39.8 × $182	=	$7,244
(c)	15.1 × $250	=	($3,775)
	Net underperformance	=	$10,235

This represented a saving to owners of $2,613 for a single voyage. It will be noted that no allowance was used on the diesel oil consumption which is normal when overperforming.

If only one speed and consumption are stated in the charterparty it would be most unusual for charterers to make an under-performance claim if the voyage was performed at a somewhat lower speed than that stated in the charterparty. Some charterparties, when giving additional speeds and consumptions, qualify the figures by stating that they are not guaranteed.

The effects of ocean currents are sometimes taken into account, depending on the voyages under consideration. For example, if repetitive transatlantic or Pacific voyages are under consideration, it would be reasonable to assume that they would cancel each other out.

However, if we take the trading pattern of a typical Panamax bulker, which could be Rotterdam—Newport News—Japan—Australia—Rotterdam, there is a definite disadvantage to shipowners. Westbound ocean passages in the Northern hemisphere are invariably adverse and the Australia—Europe portion is also generally adverse, leaving only the Japan—Australia sector as being generally favourable. This should be taken into account, especially if the vessel performs favourably on the Japan—Australia sector, but not on the other sectors.

Ocean currents can be estimated using ocean routeing charts, the weather routeing services or by relying on ships' staff to fill in a suitably designed log abstract, as mentioned in earlier chapters.

Ships' staff have to be encouraged to fill in these abstracts faithfully. In an actual case, the speed of a new vessel dropped by nearly two knots for a single day's entry without any explanation. This brought the performance into the penalty area and immediate enquiries revealed that a two knot current had been encountered without it being recorded in the log abstract then in use and was not designed to record currents.

If we take a quite simple example of how one day's wind force entered incorrectly can effect the bottom line, we can go back to the example shown in Table 10.

Say that in the entry for 12 October the wind force had been entered as force 5 instead of force 4, the revised fairweather averages would be:

> 12.40 knots   28.24 tonnes H.O.   2.11 tonnes D.O.

instead of:

> 12.44 knots   28.23 tonnes H.O.   2.11 tonnes D.O.

Although the difference appears quite small, it would add another $409 to the under-performance claim had the vessel been performing below warranted figures. This emphasises the need for extreme care when filling in abstracts, and it is all too often left to the second mate to attend to this most important task.

It is generally accepted that weather conditions calculated are on an average daily basis, even though weather conditions can change several times throughout the day. In the event of a dispute arising leading to an arbitration it is possible that deck log books could be produced and lesser periods than a day taken into account.

We must not forget that the NYPE charterparty form, as originally drawn up, does state "best grade of fuel oil". It has been established in the United States courts that charterers are responsible for the quality of bunkers supplied to any vessel they have chartered. In the event of poor quality fuel being supplied charterers should be put on notice, *not* the bunker suppliers — unless there are good reasons for doing this.

Readers attention is drawn to the BIMCO fuel quality clause, mentioned in chapter two, which covers the question of poor quality bunkers.

2. Galley fuel

Clause 20 of the NYPE charterparty form reads:

> "Fuel used by the vessel whilst off hire, also for cooking, condensing water, or for grates and stoves to be agreed as to quantity, and the cost of replacing same, to be allowed by owners".

A considerable amount of correspondence has been generated trying to define, or more appropriately update, this clause for a present-day situation. Grates and stoves are not used aboard ships these days, and many arbitrations have attempted to define what "grates and stoves" mean in current terminology.

Returning to Clause 20 it would, of course, be ideal if the quantity could be agreed at the signing of the charterparty, thus avoiding lengthy correspondence. An even better solution would be to delete the clause completely as so often happens nowadays especially if the latest revised version of the NYPE charterparty form is used.

Until quite recently, it was generally understood that the amount of fuel consumed under Clause 20 was in the region of 0.1 tonnes of diesel oil per day.

However, the Court of Appeal in a recent judgement over the *"Sounion"* case (December 1986) ruled that all fuel consumed for crew purposes must be paid for by the owners. This puts a completely different complexion on the issue and lighting, air conditioning and refrigeration must be taken into account in addition to what was referred to as "crew indulgences".

A technical approach can, of course, be taken in order to determine the amount of fuel consumed under this clause but the calculations are far from simple and some may argue not worth the effort for the rather small amounts involved.

The calculations have been further complicated by the now standard application of energy conservation measures employed in the generation of electrical power onboard modern vessels. These range from the use of blended or heavy fuel in conventional diesel alternators to steam turboalternators by which all electrical energy is supplied free when full steaming and partially when operating at reduced output (slow steaming).

The first step in applying a technical approach is to calculate the electrical load used for crew purposes which will, of course, vary from vessel to vessel depending on the capacity of the varied equipment installed.

A calculation carried out by the author taking into account loading, demand and diversity indicated that on a typical vessel the following electrical loads for crew purposes would be considered reasonable:

	kW
Galley stove	5
Condensing water	4
Lighting, etc	6
Refrigeration	4
Air conditioning/heating	7
Total	26

We know that a modern diesel alternator has a specific fuel consumption of around 215 g/kW/hour when burning diesel oil and we can easily calculate that 1 kW is equal to $215 \times 24 = 5{,}160$ g/day or 0.00516 tonnes per day. So, if our typical vessel has an electrical load of 26 kW we can say that this is equal to 0.134 tonnes of diesel oil per day or $30.4 \times 0.134 = 4$ tonnes per month.

As previously mentioned, however, the generation of electrical power using diesel alternators burning diesel oil is now rarely used except for older vessels which still have to be catered for in our calculations.

When using blended or heavy fuel the specific fuel consumption will increase due to the lower heat content of these inferior fuels. We can calculate that the daily fuel consumptions would be 0.139 and 0.142 tonnes per day respectively when using these fuels. In the case of a power take off we must use the specific fuel consumption of a modern Main Engine currently around 174 g/kW/hour or 0.109 tonnes heavy oil per day.

It should be mentioned that the actual heat used in accommodation heating systems and also in freshwater evaporators is normally part of a heat recovery system and the electrical energy used is simply for anciliary circulation duties.

We now have to convert the electrical energy into money and a summary of the various alternatives is shown in Table 12.

Table 12. Galley fuel cost table

Power generation source	Fuel in use	Cost per tonne	Fuel per day	Monthly cost $
Diesel alternator	D.O.	170	.134	692
Diesel alternator	D.O./H.O. blend	120	.139	507
Diesel alternator	H.O.	70	.142	302
Power take off	H.O.	70	.109	232
Turboalternator full steaming	—	—	—	—
Turboalternator slow steaming	D.O./H.O. blend	120	.100	365

In the case of the turboalternator slow steaming option it is assumed that the shortfall in electrical output is made up by a diesel alternator running on blended fuel. Not included in the table is the power turbine option which in a suitably designed system could be treated in similar fashion to the turboalternator scheme. Fuel costs shown were applicable at the time of writing but such is the unpredictability of the bunker market that ruling costs could be somewhat different. Even so it is comparatively easy for the reader to insert the relevant H.O. or D.O. costs into the calculation.

3. Hold cleaning

It has been found that the fuel consumed when cleaning holds is rarely apportioned correctly. Normally, it is simply included in the diesel oil consumed by the generator. By rights it should be entered separately, and not included in any performance calculation when a vessel is time chartered. The amount involved is quite small, but it can be used to offset that claimed by charterers for galley fuel. An investigation by the author revealed that the hold cleaning kW consumption per ballast voyage amounted to 4,800. If we assume a 60-day round trip the hourly rate would be 3.33 kW which with diesel oil at $170 per tonne amounts to $89 per month.

4. Tankers

Unlike bulk carriers, who generally seem to favour the NYPE charterparty form, tankers have a whole selection to choose from. Each of the oil majors have their own forms which are revised regularly so we have a situation whereby Shelltime 3 or Texacotime 2 may be the form in use.

From the performance point of view, the main difference is that the word "about" is not used in the standard versions of the forms when referring to speed and consumption. Neither is there a reference to good weather conditions — it usually being spelled out as a certain Beaufort number. Shelltime 4 has a particularly severe weather clause which reads: "excluding . . . (II) any days, noon to noon, when winds exceed force 8 on the Beaufort scale for more than 12 hours."

So the benefits normally expected when using the NYPE form, namely half knot tolerance on speed, 3% on fuel consumption, and weather conditions above force 4 excluded, cannot be taken for granted, and will not be allowed unless specifically stated.

Some of the standard tanker charterparty forms do, however, have a clause allowing owners to receive a bonus in the event of the vessel overperforming against the warranted speed and consumption.

As far as is known there is no agreed method of calculating an underperformance claim when a proportion of the voyage is spent in heavy weather; always assuming a separate weather clause has been added to the charterparty. On occasions it has been known for it to be spelled out on the lines described for the NYPE form which, readers may recall, is based on a principle that if a vessel underperforms in good weather it will similarly underperform in heavy weather. However, by not using this principle a difference does arise when making the calculations comparing NYPE with a standard tanker form.

If we go back to the previous example for the Panamax bulker the performance figures under the NYPE calculation was: 12.44 knots on 28.23/2.11 in good weather.

We will increase the charterparty speed to 13 knots to enable a quantitative result to be given, because the vessel was overperforming when applying the original NYPE calculation.

NYPE revised calculation

$$\text{Fairweather time} = \frac{5,000}{12.44} = 401.93 \text{ hours}$$

$$\text{C/P time} = \frac{5,000}{13} \times 384.61 \text{ hours}$$

    (1) Time lost = 17.32 hours

$$\text{Fairweather H.O.} = \frac{401.93}{24} \times 28.23 = 472.77 \text{ tonnes}$$

$$\text{C/P H.O.} = \frac{5,000}{13 \times 24} \times 27.43 \times 1.03 = 452.77 \text{ tonnes}$$

    (2) Excess H.O. = 20 tonnes

$$\text{Fairweather D.O.} = \frac{401.93}{24} \times 2.11 \times 1.03 = 36.40 \text{ tonnes}$$

$$\text{C/P D.O.} = \frac{5,000}{13 \times 24} \times 2.03 \times 1.03 = 33.50 \text{ tonnes}$$

    (3) Excess D.O. = 2.9 tonnes

(a)	Excess hire 17.32 × 215.45	=	$3,731.59
	Commission	=	($139.93)
(b)	Excess H.O. 20 × $70	=	$1,400.00
(c)	Excess D.O. 2.9 × $170	=	$493
	Total	=	$5,484.66 underperformed

Typical tanker calculation

$$\text{Fairweather time} = \frac{2,976}{12.44} = 239.30 \text{ hours (from log)}$$

$$\text{C/P time} = \frac{2,976}{13} = 228.92 \text{ hours}$$

    (1) Time lost = 10.31 hours

$$\text{Fairweather H.O.} = \frac{239.3}{24} \times 28.23 = 281.5 \text{ tonnes (from log)}$$

$$\text{C/P H.O.} = \frac{239.3}{24} \times 27.43 = 273.5 \text{ tonnes}$$

    (2) Excess H.O. = 8 tonnes

$$\text{Fairweather D.O.} = \frac{239.3}{24} \times 2.11 = 23.04 \text{ tonnes}$$

$$\text{C/P D.O.} = \frac{239.3}{24} \times 2.03 = 20.24 \text{ tonnes}$$

    (3) Excess D.O. = 2.80 tonnes

(a)	Excess hire 10.31 × 215.45	=	$2,221.29
	Commission	=	($83.30)
(b)	Excess H.O. 8 × $70	=	$560.00
(c)	Excess D.O. 2.8 × $170	=	$476
	Total	=	$3,173.99 underperformed

    This illustrates that, by and large, a typical tanker charterparty underperformance claim when taking heavy weather into account is not as disadvantageous to owners as the NYPE heavy weather calculation. Some charterers actually work out the result according to the tanker calculation and then multiply it by a Total Mileage Factor in the ratio:

$$\frac{\text{Total mileage}}{\text{Fairweather mileage}} = \frac{5,000}{2,976} = 1.68$$

This will then give a result according to the NYPE route — not forgetting that the 3% allowance is not given in a typical tanker claim.

5. Summary of Chapter Five

Underperformance claims should be thoroughly checked by owners using the points made in this chapter. The risk in accepting an all weather clause should be fully appreciated by all those involved. It is to be hoped that the question of galley fuel is now understood from a technical viewpoint. Tankers have their own charterparty forms, each having quite different clauses. Owners should realise what the significance of some of these clauses mean with respect to underperformance and any subsequent claim.

CHAPTER SIX

Computer application

1. Introduction

In previous chapters we have read about performance monitoring and charterparty performance, both of which are ideal candidates for computer programs. There are many programs available on the market at varying levels of sophistication and, of course, cost. When deciding what program is best suited, it should be remembered that whatever is inputted must, of course, be recorded on board. This will probably mean redesigning the logs and abstracts to make them computer friendly.

Another fundamental decision will be whether to provide a computer aboard the vessel or to have it based in the shipowner's office. A lot will depend on whether a computer is already provided aboard to perform other functions such as planned maintenance, inventory control or other subjects.

Most vessels already have a dedicated computer for carrying out loading calculations and one solution, especially with older vessels requiring this instrument to be renewed, is to replace it with a standard computer therefore giving access to a host of programs.

Probably all shipowners' offices have computers performing a variety of tasks related to payroll and accounts; the additional programs covered in this chapter would cause no problems.

2. Computer required

The author has favoured a micro or personal computer for carrying out tasks related to ship performance, and this has been found adequate for dealing with over 30 ships on a hard disc. It covers such subjects as voyage printouts which replaces the abstract, charterparty calculations and main engine performance with many other programs possible depending on the extent of the raw data inputted. It would be possible to use a larger mini computer or even a mainframe, but to keep the cost at a reasonable level the choice of a quite powerful personal computer is considered the best solution.

3. Data input

The computer can only analyse whatever data is inputted, and it is important to spend some time deciding the exact amount of data necessary. This will depend on the shipowner's requirements; for example, does he have the trained staff necessary to carry out this duty?

Regarding what might be called a typical abstract, we can assume all the normal information is required; this will include distance steamed, time, fuel consumption, kW loading and the like. In order to carry out a more detailed analysis of a vessel's performance, we should also consider other important parameters, most of which were touched on in chapter three.

When considering computer application only digital inputs are suitable in the type of computer we are considering, so everything must be reduced to a digital input and any analogue descriptions which may have been previously used must be converted to digital. This was the main purpose of converting weather conditions to digital as discussed in chapter three.

The format eventually decided on by the author is illustrated in Figure 18:

Fig. 18. Passage summary — computer printout

```
Ship Name

NOLA                              ROTTERDAM                  0012L 10/05/85 28/05/85

Mask :      Beaufort No:  0 Current Speed:  0 Swell Force:  0 Pitch/Roll:  0
```

Page 1 Date	Latitude Deg Min	Long°ude Deg Min	C/S Deg	Wind S Bn	Swell S For	Curr. S Spd	Dist 0.G	Dist Log	Pitc Roll	Deck Wet	Time Hours	Speed Knots	RPM	Slip %
10/05/85	27 29 N	87 35 W	V	2 5	2 2	0 0	138	138	2	1	12.3	11.22	55.7	5.38
11/05/85	24 30 N	83 23 W	128	1 4	1 2	4 15	290	263	2	1	24.0	12.08	58.0	2.14
12/05/85	26 51 N	79 37 W	V	8 3	6 2	8 15	308	260	2	1	24.0	12.83	57.2	-5.39
13/05/85	31 0 N	78 19 W	V	1 5	3 2	8 10	259	251	4	1	23.0	11.26	57.4	7.84
14/05/85	34 59 N	74 35 W	V	5 3	5 2	8 20	309	263	4	1	24.0	12.88	57.6	-5.00
15/05/85	37 52 N	70 0 W	V	8 4	8 5	8 18	289	247	4	2	24.0	12.04	58.7	3.64
16/05/85	38 56 N	63 54 W	V	8 6	8 5	8 25	298	227	4	2	24.0	12.42	57.9	-0.74
17/05/85	39 34 N	58 22 W	V	5 5	7 5	0 0	266	236	4	1	23.0	11.57	57.4	5.35
18/05/85	40 14 N	53 18 W	V	5 8	5 8	0 0	238	237	5	4	24.0	9.92	57.5	18.99
19/05/85	40 42 N	48 15 W	V	5 8	5 8	0 0	244	233	5	4	24.0	10.17	57.7	17.23
20/05/85	42 26 N	43 11 W	65	5 6	5 6	0 0	252	242	5	2	23.0	10.96	57.4	10.33
21/05/85	44 22 N	37 32 W	65	3 4	5 5	8 13	273	251	4	1	24.0	11.38	57.4	6.75
22/05/85	45 48 N	32 3 W	70	2 4	2 2	8 5	248	237	4	1	23.0	10.78	57.4	11.76
23/05/85	47 14 N	26 12 W	70	4 5	2 3	8 10	257	234	4	1	24.0	10.71	57.4	12.37
24/05/85	48 15 N	20 15 W	76	4 5	4 5	8 10	249	227	4	1	23.0	10.83	57.4	11.40
25/05/85	49 9 N	14 7 W	V	7 6	7 7	0 0	249	247	6	2	24.0	10.38	57.4	15.53
26/05/85	49 35 N	8 3 W	84	5 6	5 8	0 0	240	235	6	3	23.0	10.43	57.6	14.90
27/05/85	50 2 N	1 37 W	V	5 4	5 4	0 0	252	189	4	1	24.0	10.50	57.5	14.22
28/05/85	51 57 N	3 29 E	V	5 4	5 4	0 0	206	200	3	1	18.3	11.26	57.4	7.88
Totals/Averages							4865	4417			432.6	11.25	57.5	8.14

Page 2 Date	Engine	Cargo	Tank	T.A.	D.A.	Total	DA Pr	DA Ot	M.Eng	Total	T.A.	D.A.	Total
	Heavy Fuel Oil (Tonnes)						Diesel Oil (Tonnes)				Electrical Load (kw)		
10/05/85	7.55	0.00	0.00	0.50	0.00	8.05	1.20	0.00	0.00	1.20	0.0	400.0	400.0
11/05/85	15.10	0.00	0.00	1.20	1.60	17.90	1.00	0.00	0.00	1.00	0.0	410.0	410.0
12/05/85	14.35	0.00	0.00	0.85	1.40	16.60	0.90	0.00	0.00	0.90	0.0	400.0	400.0
13/05/85	14.30	0.00	0.00	1.00	1.20	16.50	1.00	0.00	0.00	1.00	0.0	400.0	400.0
14/05/85	15.00	0.00	0.00	1.00	1.60	17.60	1.10	0.00	0.00	1.10	0.0	400.0	400.0
15/05/85	16.60	0.00	0.00	1.00	1.40	19.00	1.00	0.00	0.00	1.00	0.0	410.0	410.0
16/05/85	16.40	0.00	0.00	1.00	1.50	18.90	1.00	0.00	0.00	1.00	0.0	410.0	410.0
17/05/85	14.20	0.00	0.00	0.90	1.45	16.55	0.85	0.00	0.00	0.85	0.0	410.0	410.0
18/05/85	16.60	0.00	0.00	0.90	1.55	19.05	0.95	0.00	0.00	0.95	0.0	400.0	400.0
19/05/85	17.35	0.00	0.00	1.00	1.40	19.75	0.90	0.00	0.00	0.90	0.0	400.0	400.0
20/05/85	14.50	0.00	0.00	0.90	1.30	16.70	1.00	0.00	0.00	1.00	0.0	390.0	390.0
21/05/85	14.30	0.00	0.00	1.00	1.40	16.70	0.90	0.00	0.00	0.90	0.0	390.0	390.0
22/05/85	13.70	0.00	0.00	0.90	1.50	16.10	0.90	0.00	0.00	0.90	0.0	390.0	390.0
23/05/85	16.80	0.00	0.00	0.95	1.58	19.33	1.06	0.00	0.00	1.06	0.0	380.0	380.0
24/05/85	14.05	0.00	0.00	1.20	1.40	16.65	0.85	0.00	0.00	0.85	0.0	370.0	370.0
25/05/85	16.70	0.00	0.00	1.30	1.40	19.40	0.90	0.00	0.00	0.90	0.0	360.0	360.0
26/05/85	15.70	0.00	0.00	1.10	1.30	18.10	0.90	0.00	0.00	0.90	0.0	360.0	360.0
27/05/85	15.20	0.00	0.00	1.20	1.35	17.75	0.90	0.00	0.00	0.90	0.0	360.0	360.0
28/05/85	12.75	0.00	0.00	2.50	1.05	16.30	0.65	0.00	0.00	0.65	0.0	360.0	360.0
Totals	281.15	0.00	0.00	20.40	25.38	326.93	17.96	0.00	0.00	17.96			
Averages	15.60	0.00	0.00	1.13	1.41	18.14	1.00	0.00	0.00	1.00	0.0	389.5	389.5

Starting with the heading we have ship's name, sailing port and arrival port, with dates and voyage number.

Each day's entry comprises:

Page one
Date
Latitude
Longitude
Course steered
Wind force/direction (sector)
Swell force/direction (sector)
Current speed/direction (sector)
Distance — over ground
Distance — log
Pitch/roll
Deck wetness
Time
Speed (calculated by program)
RPM
Slip (calculated by program)

Page two

Heavy oil consumption	Main Engine
	Cargo heating
	Tank cleaning
	Turbo-alternator
	Diesel alternator
	Total (calculated)
Diesel oil consumption	Diesel alternator (propulsion)
	Diesel alternator (other)
	Main Engine
	Total (calculated)
Electrical load kW	Turbo-alternator
	Diesel alternator
	Total (calculated)

It will be seen that for each day's entry there are a maximum of 27 entries if all possibilities are used. This is extremely unlikely, and the average in general use is around 21.

From this wealth of information we can cater for a comprehensive range of programs which will be discussed later in the chapter. The only important parameter not allowed for in this program is brake horsepower, and for vessels fitted with a torsionmeter, a slight modification would be required to enable the torque reading to be entered.

A brief comment will be made on each of the entries giving the reason why they have been included.

The date, of course, requires no explanation, and latitude and longitude have been included for possible expansion of the menu of programs which will eventually include voyage strategy. It is possible to divide the world into sectors and, with the help of routeing charts, we can have a historical weather database available on a computer program. We must also consider electronic charts which are gairing in popularity and, of course, a vessel's position is now displayed on demand with the universal use of satellite positioning devices with a good possibility of direct input into the computer. For these reasons it was deemed necessary to include latitude and longitude. Also, we must not forget it gives a guide to the route taken for post-voyage analysis purposes. Course steered could also be placed in the category of latitude and longitude, namely for future possibilities related to voyage strategy.

Wind, swell and current need no additional comment except, perhaps, that they are in numerically ascending order of magnitude both in force and direction, which would also apply to the weather related effects of pitch/roll and deck wetness.

Distance over ground and by log are nowadays from instruments usually very accurate in the case of over ground and, dependent on the sophistication of the method employed, the recording of log distance.

RPM is usually very accurate and speed over ground and slip are calculated from the data inputs relating to distance, time and RPM with the propeller constant already fed into the program.

Heavy fuel oil consumption has been divided into five separate consumers, namely Main Engine, cargo heating, tank (or hold) cleaning, turbo-alternator and diesel alternator. Whilst this extensive subdivision is mainly for tanker operation, it can be useful for identifying hold cleaning operations on bulkers in certain instances. If flashing a boiler is necessary to provide steam for the turbo-alternator the fuel consumed would, of course, be entered in the turbo-alternator column and, although this method of electricity generation is normally not efficient, a provision has been made for it.

Diesel oil consumption has been split into three consumers — namely diesel alternator propulsion, diesel alternator other and Main Engine. Diesel oil consumed entered in the other column would include tank and hold cleaning and also tank heating if this split was necessary for charterparty purposes. It is unlikely that the Main Engine would consume diesel oil, but a provision has been made in case emergency operation requires its use.

Finally, there is electrical load divided between diesel and turbo-alternators with the total kW load shown.

These, then, are the daily inputs which form the basis of the various programs. There are other inputs relating to predicted performance, propeller constant, displacement curve and Main Engine parameters, but these will be discussed later.

4. Passage summary

The first program is simply a printout of all the daily data for whatever voyage by whatever ship is selected as shown in Figure 18. Included in the program is a weather mask which permits Beaufort number limits to be pre-selected and also current speed limits and swell intensity limits, so that a realistic performance in simulated good weather is available on demand.

The second program is a variant of the first, again on a daily basis, but concentrating on the speed aspect as shown in Figure 19. It will be seen that speed over ground, log speed and speed over the ground with current effect are shown. This later feature was developed to suit the requirements of a charterer who used this speed when calculating charterparty performance.

Current speed is indicated in tenths of a knot and the sectors are as described in Figure 12. It will be noted that the log speed recorded is apparently inaccurate, and experience has shown that this is a feature throughout the fleet. It also brings into question the reliability of the figures entered for current speed.

Fig. 19. **Speed summary — computer printout**

```
NOLA                       ROTTERDAM                  0012L 10/05/85 28/05/85

Mask :      Beaufort No:  O Current Speed:  O Swell Force:  O Pitch/Roll:  O
-----------------------------------------------------------------------------
Date      Dist  Time  BN   Current    Current    --------- Speed Knots ---------
                           Sector     Speed*10   Over Grnd     Log     OG+Current
```

Date	Dist	Time	BN	Current Sector	Current Speed*10	Over Grnd	Log	OG+Current
10/05/85	138	12.3	5	Variable	0	11.220	11.220	11.220
11/05/85	290	24.0	4	SECTOR 4	15	12.083	10.958	12.083
12/05/85	308	24.0	3	SECTOR 8	15	12.833	10.833	14.333
13/05/85	259	23.0	5	SECTOR 8	10	11.261	10.913	12.261
14/05/85	309	24.0	3	SECTOR 8	20	12.875	10.958	14.875
15/05/85	289	24.0	4	SECTOR 8	18	12.042	10.292	13.842
16/05/85	298	24.0	6	SECTOR 8	25	12.417	9.458	14.917
17/05/85	266	23.0	5	Variable	0	11.565	10.261	11.565
18/05/85	238	24.0	8	Variable	0	9.917	9.875	9.917
19/05/85	244	24.0	8	Variable	0	10.167	9.708	10.167
20/05/85	252	23.0	6	Variable	0	10.957	10.522	10.957
21/05/85	273	24.0	4	SECTOR 8	13	11.375	10.458	12.675
22/05/85	248	23.0	4	SECTOR 8	5	10.783	10.304	11.283
23/05/85	257	24.0	5	SECTOR 8	10	10.708	9.750	11.708
24/05/85	249	23.0	5	SECTOR 8	10	10.826	9.870	11.826
25/05/85	249	24.0	6	Variable	0	10.375	10.292	10.375
26/05/85	240	23.0	6	Variable	0	10.435	10.217	10.435
27/05/85	252	24.0	4	Variable	0	10.500	7.875	10.500
28/05/85	206	18.3	4	Variable	0	11.257	10.929	11.257

```
-----------------------------------------------------------------------------
Average Speeds (knots)                              11.246     10.210    11.939
-----------------------------------------------------------------------------
```

The third program is shown in Figure 20.

Fig. 20. **Performance summary — computer printout**

```
NOLA                       ROTTERDAM                  0012L 10/05/85 28/05/85

Mask :      Beaufort No:  0 Current Speed:  0 Swell Force:  0 Pitch/Roll:  0
-----------------------------------------------------------------------------
```

Date	Dist miles	Time hours	BN	Wind sect	Speed knots	RPM	Slip %	Load %	Speed diff	Fuel diff
10/05/85	138	12.3	5	2	11.22	55.7	5.38	112.77	-0.57	2.04
11/05/85	290	24.0	4	1	12.08	58.0	2.14	102.37	0.25	-2.09
12/05/85	308	24.0	3	8	12.83	57.2	-5.39	101.43	1.09	-8.46
13/05/85	259	23.0	5	1	11.26	57.4	7.84	104.37	-0.55	2.12
14/05/85	309	24.0	3	5	12.88	57.6	-5.00	103.83	1.06	-8.11
15/05/85	289	24.0	4	8	12.04	58.7	3.64	108.57	0.03	-0.25
16/05/85	298	24.0	6	8	12.42	57.9	-0.74	111.76	0.43	-3.43
17/05/85	266	23.0	5	5	11.57	57.4	5.35	103.64	-0.23	1.18
18/05/85	238	24.0	8	5	9.92	57.5	18.99	115.51	-2.10	7.46
19/05/85	244	24.0	8	5	10.17	57.7	17.23	119.47	-1.94	7.53
20/05/85	252	23.0	6	5	10.96	57.4	10.33	105.83	-0.88	3.16
21/05/85	273	24.0	4	3	11.38	57.3	6.75	100.55	-0.36	1.19
22/05/85	248	23.0	4	2	10.78	57.4	11.76	99.99	-0.95	2.80
23/05/85	257	24.0	5	4	10.71	57.4	12.37	117.51	-1.33	5.50
24/05/85	249	23.0	5	4	10.83	57.4	11.40	102.55	-0.95	3.04
25/05/85	249	24.0	6	7	10.38	57.7	15.53	115.00	-1.65	6.31
26/05/85	240	23.0	6	5	10.43	57.6	14.90	113.40	-1.55	5.83
27/05/85	252	24.0	4	5	10.50	57.5	14.22	105.76	-1.34	4.47
28/05/85	206	18.3	4	5	11.26	57.4	7.88	116.36	-0.77	3.93

```
-----------------------------------------------------------------------------
FC 150154  4865  432.6               11.25  57.5  8.14  108.49   -0.65   2.84
-----------------------------------------------------------------------------
```

In this program certain calculations are made based on information which has previously been inputted, and this is what might be called the expected performance. This is shown in Table 13.

Table 13.   Expected performance — computer input

Output	RPM	M.E. H.O.	Boiler H.O.	Diesel gen. H.O.	Diesel gen. D.O.	Load speed	Ballast speed
100%	78	36	—	—	—	15.00	15.75
80%	74	30	—	0.5	0.6	14.25	15.25
65%	69	24	—	0.7	0.8	13.75	14.50
50%	64	19	1	1.0	1.1	13.00	13.75
35%	57	14	1	1.2	1.3	12.00	13.00
25%	53	11	1	1.2	1.3	10.75	11.75

If we consider the load percentage column this is, in fact, the absorbed load based on the assumption that the Main Engine fuel consumption and revolutions as per the above input are 100%

It will be noted that the Main Engine consumption inputs are rounded off to suit charterparty consumptions which are normally given in whole numbers. Rather than use more accurate fuel consumptions based on cube law revolutions and corrected test bed specific fuel consumption, it is felt that using charterparty figures avoids duplication.

The overload figure can then be used purely as a trend indicator, rather than an accurate reflection of the absorbed load. It is interesting to note that in this particular case a derated engine is fitted and the absorbed load indicates around 108% instead of the expected 100% for a non-derated engine. For non-derated engines it is considered dangerous to operate above 115% absorbed load for extended periods, especially if running at a high output. Deviations from the expected performance, with respect to speed and consumption, are shown in the two final columns, and do not require special comment. The weather screening option is available for all these programs.

5. Database

Each voyage is summarised in Figure 21.

Fig. 21.   Database — computer printout

```
Mask :-  Beaufort No  0: Current Speed  0: Swell Force  0: Pitch/Roll  0
------------------------------------------------------------------------
Voyage No.  Distance  M.E:- t/d    D.A :- Kw     T.A :- Kw   Slip %      Displ.
Index No.   Time hrs        RPM    HO/DO t/d       HO t/d    Speed Diff  Trim
            Speed         Load %   gm/bhp/hr      Save t/d   Fuel Diff   Fuel Coeff
------------------------------------------------------------------------

NOLA                      ROTTERDAM                10/05/85   28/05/85

0012L        4865         15.60       389.47       0.00         8.14      66821
  22       432.60         57.48  1.41  1.00         1.13        -0.65       0.00
            11.25        108.49       191.89      -1.13         2.84     150154
```

The first three columns are self-explanatory, and do not need any comment. The fourth column relates to the performance of the diesel alternator giving average load, also daily consumption of heavy oil and/or diesel oil together with the specific fuel consumption. The next column shows the performance of the turbo-alternator in similar fashion as the diesel alternator but, additionally, shows the benefit of the turbo-alternator, expressed in tonnes per day fuel saved for the "free" kW produced. In this particular example the turbo-alternator was not in use due to a defect in the economiser necessitating the boiler being flashed to support the steam load. An example of how this works will be shown later. In the sixth column deviations from the expected performance are shown as well as the slip. In the final column the vessel's displacement (not deadweight) is shown, as is the trim based on sailing and arrival draughts, also the fuel coefficient using the mean draught.

The database can be printed for single or multiple voyages with a selection option for loaded or ballast, or any other combination of voyages at will.

6. Graphic illustration of trend analysis

A selection of graphic illustrations are available, all on an elapsed time basis. Whereas the previous programs have considered voyage performance in isolation the next set of programs indicate long-term trends for selected voyages either loaded, ballast or both, in chronological order.

These programs will process raw passage data for whatever voyages have been specified and make corrections for displacement, revolutions and weather. It is also possible additionally to screen the data using the weather mask and current speed.

The corrected data will be displayed graphically with speed or fuel consumption plotted against elapsed time. Voyage number and start date of voyage will be indicated on the time axis and events such as dry dockings, bottom scrubs and propeller polishings shown. An illustration of the corrected raw data is shown in Figure 22.

From these corrected raw data a five-part moving average will then produce smoothed speed or fuel consumption in similar manner as shown in Figure 23.

Average values for speed and consumption will be calculated for each voyage, and these values will be smoothed using a three-part moving average.

To complete this section of the program an analysis of absorbed engine loading with RPM reference points will be shown graphically as illustrated in Figure 25.

This, more or less, covers the technical aspects of computerised performance monitoring which has a distinct advantage over a manual system with respect to the speed of the calculations and the possibility of forecasting trends.

7. Charterparty calculations

The raw data inputted into the passage summary are normally adequate for dealing with most charterparty performance claims.

We have seen how weather screening possibilities can cover all known variations so far met with when dealing with underperformance claims. Such a comprehensive weather description used in the computerised passage data program is not normally required for charterparty calculations. It is, however, very important that the data must be identical. It is no use working out a claim submitted by charterers based on their own abstracts and then using owners' abstracts to challenge the claim unless both sets of figures are identical on the important issues of distance, time, fuel and weather.

Charterers abstract forms may, possibly, be laid out quite differently to the computerised passage summary described in this chapter; in all probability the charterers form will not contain as much weather information, neither will it split fuel up into the number of consumers we have used in the computerised version.

Fig. 22.   Unsmoothed trend graphs

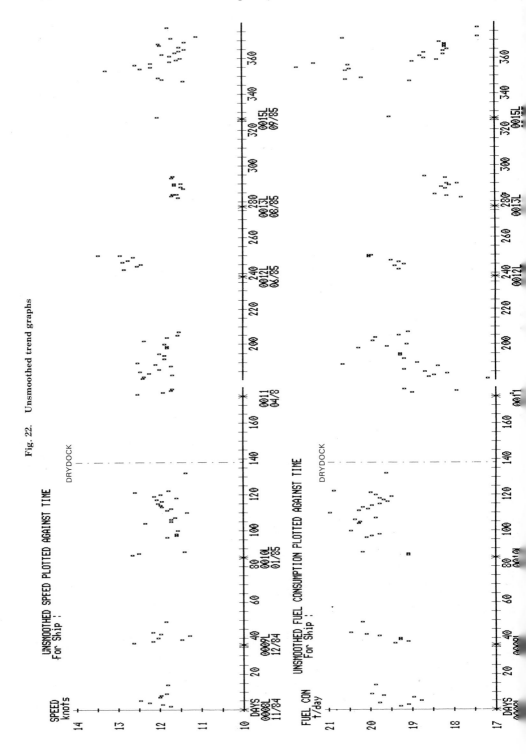

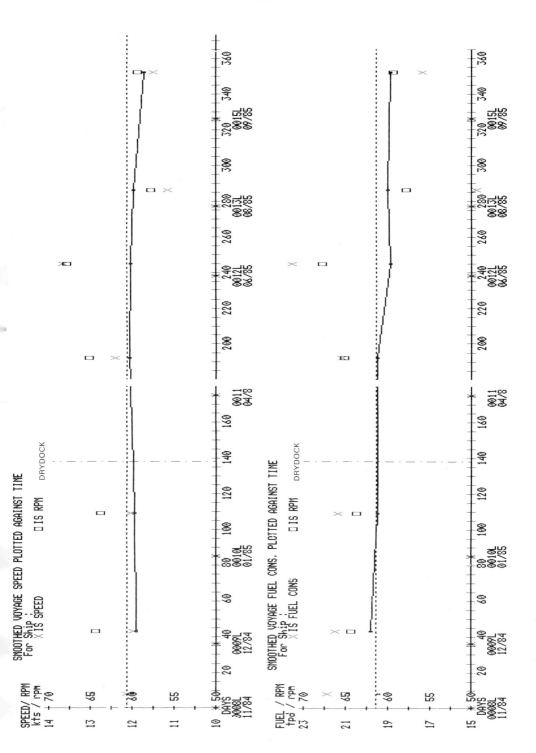

Fig. 24.   Regression analysis graphs

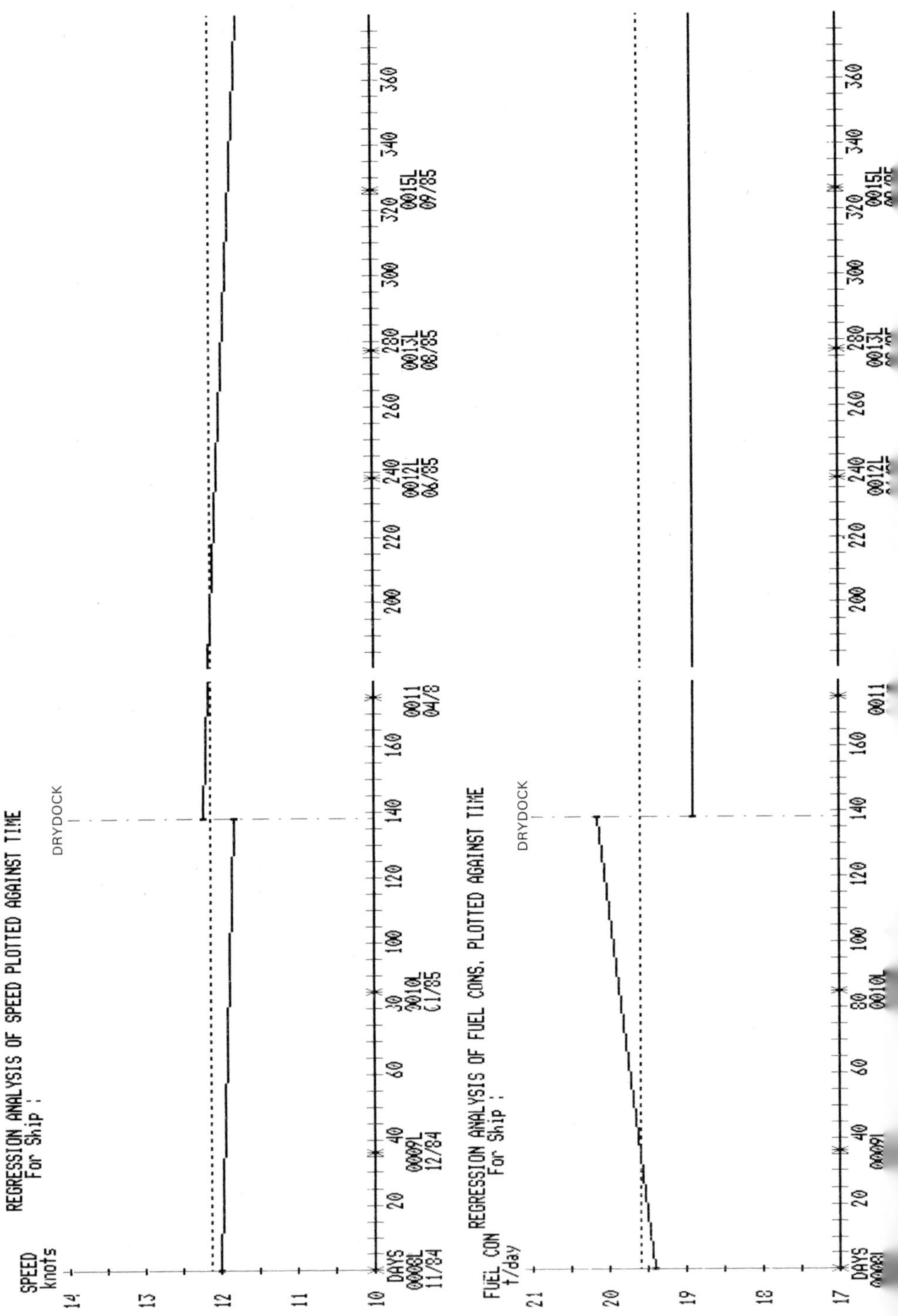

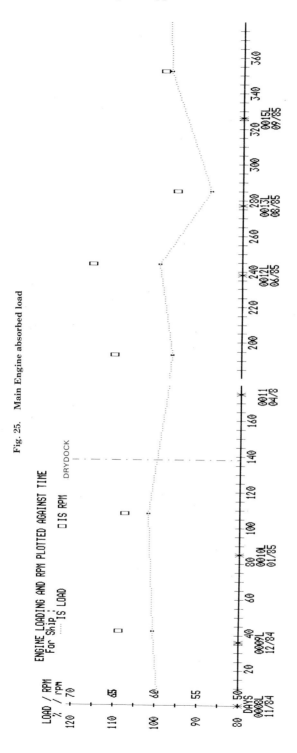

Fig. 25. Main Engine absorbed load

Ship's staff must be encouraged to complete both owners' and charterers' abstracts using identical figures, and additional information, not catered for in the charterers' abstract, should be entered in the remarks column. This would generally relate to adverse currents, reductions in speed for fog or to prevent damage to the vessel in heavy weather. Fuel used for what might be called charterers' activities must also be separated from essential propulsion fuel. This would include cargo heating and tank or hold cleaning.

It is also important that charterers' steaming instructions are clearly indicated in the remarks column, both when starting the voyage and at each change in instructions received thereafter. This would include instructions to fit slow steaming fuel valve nozzles.

The computer program has a display in which all the various fuel consumers are clearly listed and can be included or omitted by the user. For example, Main Engine and diesel generator fuel would obviously be included in the performance calculations, whereas tank cleaning and tank heating would, in all probability, be omitted.

There is also an option for selecting NYPE calculations or typical tanker calculations as fully described in the previous chapter. Weather screening is included too in an option which allows Beaufort number, current speed, swell force and pitch/roll to be chosen at will.

In a typical charterparty only Beaufort number would normally be taken into the calculation, but unexplained inferior performances could be identified and perhaps negotiated with the charterer or at an arbitration. In the event of the charterer choosing to use a weather routeing service to challenge the recorded weather conditions, the weather section of the program would prove to be very useful should such a challenge be unfounded. It should be mentioned that weather routeing services very often route vessels away from heavy swells, as they consider their effect more devastating to a vessel's performance than a moderate wind force.

The program also allows a voyage to be split into five periods, each having different speed settings — should this have been asked for by the charterer.

When calculating the underperformance, or overperformance in certain cases, we must first input the voyage reference number from the menu of voyages displayed on the screen. If there were no alterations in steaming orders, the start and finish dates of the chosen voyage will not require adjustment. If charterers have instructed the vessel to speed up or slow down, the date of the first change is inputted and the program automatically rearranges the voyage into the requisite number of periods.

Next to be inputted is the charterparty speed. This can be with or without the ½ knot tolerance depending on the terms of the charter. Charterparty heavy oil consumption and diesel oil consumption are likewise entered using whatever allowance is applicable. Fuel costs of both heavy oil and diesel oil are also entered.

Daily charter freight rate, deadweight and address commission are also inputted. If the vessel is time chartered at a daily hire rate rather than a daily freight rate, it can be easily converted as shown:

$$\text{Daily Freight Rate} = \frac{\text{Daily Hire Rate} \times 30.4}{\text{Deadweight}}$$

If thought necessary, the address commission can be deducted from the daily hire rate and the address commission inputted as zero. This would not, however, be suitable if an overperformance clause was included in the charterparty terms.

The weather mask limit is then inputted, as is the charterparty form to be used, either NYPE or tanker. The various fuel consumptions to be taken into account are also selected, as previously mentioned. We can then display and print out the calculation. Should we decide that the end result is not beneficial and no allowances have been used, we can add these allowances before printing out the final version.

As an example, we will show a calculation for the voyage used in section 3 of this chapter, which readers may recall was a New Orleans—Rotterdam laden passage on a Panamax bulker.

This is shown below in Figure 26.

Fig. 26.   **Charterparty claim all weather: New Orleans—Rotterdam**

```
CHARTER PERFORMANCE  :   VOYAGE - 0012L         10/05/85 to 28/05/85
     SUMMARY         :   NOLA                      to ROTTERDAM
-------------------------------------------------------------------------

     Date From        Date To        Speed         Heavy Oil      Diesel Oil
                                      knots         tonnes/day     tonnes/day

 1   10/05/85         28/05/85        12.00          16.260          1.320



----------------------------------------------------------------------------

Heavy Fuel Oil Price ( $ / tonne) .......     70.0
Diesel      Oil Price ( $ / tonne) .......    170.0
Tonnage   (tonnes) ......................    63650
Charter Rate   ( $ /tonne/month) ........     3.000
Address   ( % ) .........................      3.70

----------------------------------------------------------------------------

Charter Party Type is ................... NYPE
Speed Type is ........................... Speed over ground

----------------------------------------------------------------------------


NOLA                     ROTTERDAM              0012L 10/05/85 28/05/85
Mask :     Beaufort No: 0 Current Speed: 0 Swell Force: 0 Pitch/Roll: 0
----------------------------------------------------------------------------
     Leg 1 of 1                     Period : 10/05/85 - 28/05/85

                                    All Weather         After Mask
Total Distance (miles) .................     4865.0           4865.0
Total Steaming Time (hours) ............     432.60           432.60
Average Speed (knots) ..................      11.25            11.25
Average RPM ............................      57.48            57.48
Average Fuel Consumption (tonnes/day) ...     18.14            18.14
Average Diesel Cons. (tonnes/day) .......      1.00             1.00

Excess Hire   ( US $ ) ..................                    -6851.16
Excess Heavy Oil Costs ( US $ ) .........                    -3658.22
Excess Diesel Costs ( US $ ) ............                      737.45

Charter Performance ( US $ ) ............                    -9771.92

Cumulative Charter Performance ( $ ) ....                    -9771.92

----------------------------------------------------------------------------
```

The charterparty performance figures were 12 knots laden on 16 l.t. heavy oil and 1.3 l.t. diesel oil which have been converted to metric tonnes to tie up with the abstract figures. It can be seen that an underperformance claim of $9,771.92 would be forthcoming from charterers if no allowances were taken into account. We can now re-calculate the underperformance using a ½ knot on speed and 3% on heavy oil and diesel.

The result is shown in Figure 27.

Fig. 27. Charterparty claim fairweather: New Orleans—Rotterdam

```
CHARTER PERFORMANCE  :   VOYAGE - 0012L        10/05/85 to 28/05/85
    SUMMARY          :   NOLA                      to ROTTERDAM
-----------------------------------------------------------------------------

      Date From        Date To       Speed       Heavy Oil       Diesel Oil
                                     knots      tonnes/day      tonnes/day

  1    10/05/85        28/05/85      11.50         16.740          1.360

-----------------------------------------------------------------------------

Heavy Fuel Oil Price ( $ / tonne) .......      70.0
Diesel     Oil Price ( $ / tonne) .......     170.0
Tonnage  (tonnes) .....................     63650
Charter Rate  ( $ /tonne/month) .........      3.000
Address  ( % ) ......................      3.70

-----------------------------------------------------------------------------

Charter Party Type is ................... NYPE
Speed Type is ........................... Speed over ground

-----------------------------------------------------------------------------


NOLA                      ROTTERDAM                  0012L 10/05/85 28/05/85
Mask :      Beaufort No:  5 Current Speed:  0 Swell Force:  0 Pitch/Roll:  0
-----------------------------------------------------------------------------
      Leg 1 of 1                        Period : 10/05/85 - 28/05/85

                                            All Weather      After Mask
Total Distance (miles) ..................      4865.0          2175.0
Total Steaming Time (hours) .............      432.60          185.30
Average Speed (knots) ...................      11.25           11.74
Average RPM .............................      57.48           57.64
Average Fuel Consumption (tonnes/day) ...      18.14           17.87
Average Diesel Cons. (tonnes/day) .......      1.00            0.95

Excess Hire  ( US $ ) ...................                      2159.39
Excess Heavy Oil Costs ( US $ ) .........                      -944.38
Excess Diesel Costs ( US $ ) ............                      1280.46

Charter Performance ( US $ ) ............                      2495.47

Cumulative Charter Performance ( $ ) ....                      2495.47

-----------------------------------------------------------------------------
```

The charterparty speed now becomes 11.5 knots and the heavy oil 16.74 tonnes, and the diesel oil 1.36 tonnes. The weather mask chosen was a limit of Beaufort 5 which means that the program will reject all Beaufort readings of five and above. The complete turn round in performance will be noted. It now shows an overperformance of $2,495.47 when using the fuel allowance and weather mask. Actually, this will not result in a claim by owners against charterers for overperformance, as it is not allowed to use the allowances and then claim for underperformance. The result would therefore be that the voyage was performed within the warranted performance with no claim on either side involved. In fact, cases have been known of charterers grouping voyages together choosing all those at say minimum power setting and all those at various other settings that they had elected to sail the vessel at.

In this case, the rather good result obtained when using the weather mask and fuel tolerances would improve voyages with poorer results. The computer program is arranged to deal with this groupage option, should it ever be exercised.

We can also show the computerised version of another example mentioned in chapter five. This was a voyage from Newcastle NSW — Rotterdam with all weather conditions included by virtue of the fact that "under good weather conditions" had been deleted from the relevant clause in the NYPE charterparty.

Charterers did allow the ½ knot speed tolerance and also the 3% fuel tolerance so that the charterparty performance of 11 knots on 29 M.T. H.O. plus 2.5 l.t. D.O. became 10.5 knots on 29.87 M.T. H.O. plus 2.61 M.T. D.O. It will be noted that charterers used metric tonnes for H.O. and long tons for D.O. which required converting to metric tonnes before entering into the program.

The printout without any weather mask was \$67,152.30 underperformed as shown in Figure 28.

Fig. 28. Charterparty claim all weather: NSW — Rotterdam

```
CHARTER PERFORMANCE :   VOYAGE - 0012L          26/09/85 to 21/11/85
     SUMMARY         :   NEWCASTLE                  to ROTTERDAM
---------------------------------------------------------------------------

      Date From        Date To        Speed       Heavy Oil        Diesel Oil
                                       knots      tonnes/day        tonnes/day

 1    26/09/85         21/11/85        10.50        29.870            2.610




---------------------------------------------------------------------------

Heavy Fuel Oil Price ( $ / tonne ) .......      152.9
Diesel     Oil Price ( $ / tonne ) .......      238.6
Tonnage    (tonnes) .....................     129913
Charter Rate  ( $ /tonne/month) .........      2.500
Address   ( % ) .........................      3.80


---------------------------------------------------------------------------

Charter Party Type is ................... NYPE
Speed Type is ........................... Speed over ground


---------------------------------------------------------------------------


NEWCASTLE                  ROTTERDAM                    0012L 26/09/85 21/11/85
Mask :    Beaufort No: 0 Current Speed: 0 Swell Force: 0 Pitch/Roll: 0
---------------------------------------------------------------------------
      Leg 1 of 1                          Period : 26/09/85 - 21/11/85

                                        All Weather        After Mask
Total Distance (miles) ..................   13415.0          13415.0
Total Steaming Time (hours) .............   1391.50          1391.50
Average Speed (knots) ...................     9.64             9.64
Average RPM .............................    80.21            80.21
Average Fuel Consumption (tonnes/day) ...    29.41            29.41
Average Diesel Cons. (tonnes/day) .......     2.45             2.45

Excess Hire  ( US $ ) ...................                  -48767.90
Excess Heavy Oil Costs ( US $ ) .........                  -17582.96
Excess Diesel Costs ( US $ ) ............                    -801.45

Charter Performance ( US $ ) ............                  -67152.30

Cumulative Charter Performance ( $ ) ....                  -67152.30


---------------------------------------------------------------------------
```

The printout with a Beaufort 5 weather mask was $4,867.87 overperformed as shown in Figure 29, but of course this would not change the fact that owners were obliged to meet the claim of $67,152.30.

Fig. 29. Charterparty claim fairweather: NSW—Rotterdam

```
CHARTER PERFORMANCE :   VOYAGE - 0012L        26/09/85 to 21/11/85
     SUMMARY          :   NEWCASTLE                 to ROTTERDAM
------------------------------------------------------------------------------

     Date From        Date To         Speed        Heavy Oil      Diesel Oil
                                      knots        tonnes/day     tonnes/day

1    26/09/85         21/11/85        10.50          29.870          2.610




------------------------------------------------------------------------------
Heavy Fuel Oil Price ( $ / tonne) .......     152.9
Diesel     Oil Price ( $ / tonne) .......     238.6
Tonnage    (tonnes) ....................     129913
Charter Rate  ( $ /tonne/month) ........      2.500
Address    ( % ) .......................      3.80

------------------------------------------------------------------------------


Charter Party Type is ................... NYPE
Speed Type is ..,........................ Speed over ground

------------------------------------------------------------------------------


NEWCASTLE                  ROTTERDAM              0012L 26/09/85 21/11/85
Mask :    Beaufort No:  5 Current Speed:  0 Swell Force:  0 Pitch/Roll:  0
------------------------------------------------------------------------------
     Leg 1 of 1                        Period : 26/09/85 - 21/11/85

                                      All Weather      After Mask
Total Distance (miles) ..................   13415.0         4940.0
Total Steaming Time (hours) .............   1391.50          473.00
Average Speed (knots) ...................      9.64           10.44
Average RPM .............................     80.21           80.26
Average Fuel Consumption (tonnes/day) ...     29.41           28.95
Average Diesel Cons. (tonnes/day) .......      2.45            2.47

Excess Hire  ( US $ ) ...................                  -2934.97
Excess Heavy Oil Costs ( US $ ) .........                   6206.13
Excess Diesel Costs ( US $ ) ............                   1596.71

Charter Performance ( US $ ) ............                   4867.87

Cumulative Charter Performance ( $ ) ....                   4867.87


------------------------------------------------------------------------------
```

It will be noted that only 20 days out of 58 days were performed in good weather which is typical of such a voyage.

The effect of the turbo-alternator is shown on the database printout for another voyage (Figure 30) and it will be seen that a saving of 1.71 M.T./Day arose.

Fig. 30. Database printout

```
Ship Name :

Mask :-  Beaufort No  5: Current Speed  0: Swell Force  0: Pitch/Roll  0
------------------------------------------------------------------------
Voyage No.  Distance  M.E:- t/d    D.A :- Kw   T.A :- Kw  Slip %       Displ.
Index No.   Time hrs       RPM    HO/DO t/d     HO t/d  Speed Diff    Trim
            Speed      Load %    gm/bhp/hr    Save t/d  Fuel Diff     Fuel Coeff
------------------------------------------------------------------------

MUTSURE                 HAY POINT              28/08/86   11/09/86

0026B        2289       21.72     333.75      132.75      -3.27       38632
   1       186.50       88.83  0.00  2.05       0.00      -0.14        2.06
            12.27      100.93     190.56        1.71       0.48       97266
```

We can also show the effect of swell force in another example and in this case three database printouts show that the best performance in terms of speed and Main Engine fuel consumption is when a swell force 8 limit is used as a weather mask (Figure 31).

Fig. 31. Database weather screening

```
Mask :-  Beaufort No  0: Current Speed  0: Swell Force  0: Pitch/Roll  0
------------------------------------------------------------------------
Voyage No.  Distance  M.E:- t/d    D.A :- Kw   T.A :- Kw  Slip %       Displ.
Index No.   Time hrs       RPM    HO/DO t/d     HO t/d  Speed Diff    Trim
            Speed      Load %    gm/bhp/hr    Save t/d  Fuel Diff     Fuel Coeff
------------------------------------------------------------------------

DUNKIRK                 NEWPORT NEWS           14/11/85   28/11/85

0083B        3685       29.51     264.33      173.00       8.16       64085
  24       344.00       78.18  0.00  1.93       0.00      -1.13        2.63
            10.71      100.97     227.25        2.23       5.95       66704


Mask :-  Beaufort No  5: Current Speed  0: Swell Force  0: Pitch/Roll  0
------------------------------------------------------------------------
Voyage No.  Distance  M.E:- t/d    D.A :- Kw   T.A :- Kw  Slip %       Displ.
Index No.   Time hrs       RPM    HO/DO t/d     HO t/d  Speed Diff    Trim
            Speed      Load %    gm/bhp/hr    Save t/d  Fuel Diff     Fuel Coeff
------------------------------------------------------------------------

DUNKIRK                 NEWPORT NEWS           14/11/85   28/11/85

0083B        1710       29.67     252.14      190.00       7.30       64085
  24       158.00       78.30  0.00  1.93       0.00      -1.03        2.63
            10.82      101.16     237.81        2.45       5.55       68433


Mask :-  Beaufort No  0: Current Speed  0: Swell Force  8: Pitch/Roll  0
------------------------------------------------------------------------
Voyage No.  Distance  M.E:- t/d    D.A :- Kw   T.A :- Kw  Slip %       Displ.
Index No.   Time hrs       RPM    HO/DO t/d     HO t/d  Speed Diff    Trim
            Speed      Load %    gm/bhp/hr    Save t/d  Fuel Diff     Fuel Coeff
------------------------------------------------------------------------

DUNKIRK                 NEWPORT NEWS           14/11/85   28/11/85

0083B        2453       29.44     256.50      170.00       5.53       64085
  24       222.00       78.40  0.00  1.97       0.00      -0.78        2.63
            11.05       99.84     238.43        2.19       4.19       73389
```

It will be seen that the final example is the best. Any combination of Beaufort number, current speed, swell force and pitch/roll can be chosen to illustrate what effect the weather is having on performance.

8. Main Engine performance

We have seen how absorbed load has already been taken care of in the computer program and further aspects can now be considered. Main Engine performance, unlike overall ship performance, is not normally an on line continuous function receiving contributions in the form of regular watch readings and observations as the ship performance does. The main event is probably when power diagrams are taken and the results analysed; this is usually, perhaps, once per passage or if on short voyages, once per week depending on the shipowner's practice. Naturally other readings are regularly taken which would give early indication of trouble developing and these are usually under the control of an alarm and monitoring system permanently connected to essential pressures and temperatures. It is considered a good idea to take a set of indicator cards or power diagrams when weather conditions are good and preferably when the engine is running at high output.

On vessels equipped with large economisers having extended heating surfaces it is advantageous when slow steaming to speed the engine up for half an hour per day to enable soot deposits to be cleared. During this period it is probably the most favourable time to take a set of cards, even though the engine is not fully stabilised from a thermal loading point of view. Even so, it is probably better than taking a set of cards when running at minimum power.

The essential data to be inputted into the program would be:

    Date
    RPM
    Weather conditions
    Draught
    P.Max (each cylinder)
    P.Comp (each cylinder)
    P.Ind (each cylinder)
    Exhaust temp (each cylinder)
    Fuel pump index (each cylinder)
    P.Scav
    Temp. Scav
    Pressure drop across air coolers
    Fuel oil particulars
    Cylinder oil consumption

Other inputs could be added if this was thought necessary to suit engine type and operating conditions, and any special requirements of the engine builders. Also inputted would be overhaul dates of turbo-charger and maximum wear on cylinder liners.

The various ratios between P.Max, P.Comp, P.Ind and P.Scav could be analysed on a trend basis, which would give an indication of both thermal efficiency and mechanical performance.

Impending faults resulting in turbo-charger surging for example, could be identified and cured before developing into a critical condition.

9. Cargo heating performance

Tankers carrying cargoes which require heating can consume quite large amounts of fuel oil; in extreme cases the daily fuel consumption used heating the cargo can approach that consumed by the main propulsion engine. It is considered that a useful program could be developed based on a databank of past voyages in which vessels with rather good results could be compared with the poor performers.

Even on sister ships carrying similar cargoes at similar times of the year, variances of 80% in the amount of fuel consumed have been recorded. A computer program based on many such voyages could provide a useful check for monitoring cargo heating performance.

10. Voyage strategy

At the moment around 100 ship years of raw passage data is stored on hard disc, covering many hundreds of voyages. This databank is growing at an annual rate of over 30 ship years, and it should shortly be possible to process this data in a mainframe computer.

Resulting from this exercise, some sort of strategy program is envisaged which, amongst other things, would categorise voyages on a seasonal geographical basis. This would enable a prediction to be made of weather effect on ship type based on historical data. It is well known that attempts to make up time in adverse weather is not normally fuel cost-effective and the precise cost could be predicted by the envisaged program.

It is also well known that a voyage carried out under constant revolutions is more cost-effective than a voyage carried out with two or more speed changes. On many occasions we find vessels have to speed up or slow down to meet deadlines which should have been known earlier, and the program would cater for this. The miles steamed per tonne of fuel consumed could be used to illustrate this point to ship's staff who, on occasions, appear unaware of the penalties involved in changing speed intervoyage.

A recent development concerns the decision to use either the Suez Canal or Cape route on a voyage from Europe to the Middle East or Far East. This, of course, is related to fuel costs and the decision-making process could easily be handled by the envisaged program.

Included in the program could be a bunker strategy section which would determine the amount of upliftings and bunker ports.

Using the trend analysis section of the existing program it would be possible to update the so-called expected performance figures which, when used in conjunction with the strategy program, would provide an extremely reliable guide to the expected performance at any given time.

Also in the program would be a set of comprehensive distance tables giving alternatives when either Suez or the Cape route are being considered. Although weather routeing services are available they can be complemented by the use of these tables. A classic example is whether or not to use the UNIMAK pass route and the program would give the fuel costs which could be related to increased revenue due to the shorter passage time. The expected performance of the vessel, by using the weather effect prediction, would form part of the calculation.

The program would be able to forecast the probable financial result of the voyage using updated performance figures modified for expected weather conditions. As mentioned earlier, other inputs would include bunkering strategy.

11. Communications

At the moment, raw performance data produced on the ship in the form of log and abstract entries are posted at the voyage end and, generally, arrive at the author's shipowner office a week or so later.

It is then entered into the computer and analysed with respect to performance, using both the technical and commercial programs. The actual time taken to enter the data takes, maybe, 20 minutes for an average voyage — so the task is not unduly onerous.

Several vessels have computers provided on board and submit floppy discs on a regular basis. The information on these floppy discs is then transferred to the hard disc of the office-based micro, and analysed in a similar manner. Ship's staff also have the possibility of analysing the data on board as they have been provided with all the analytical programs.

For all normal situations it would appear that, from a technical analytical point of view, these alternative procedures are satisfactory with little to choose from except for the moderate cost for the provision of a simple floppy disc computer.

It is rare for serious performance-related technical problems to develop without prior warning from the trend analysis program. This, of course, would not apply to sudden catastrophic breakdowns which occur from time to time, and it is unlikely that they would be detected by any standard computer program anyway. However, in the event of sudden hull bottom fouling developing, it would be picked up rather quickly by either the computer or the manual system described earlier.

The commercial considerations depend on the type of charterparty in use and future business contemplated. For example, if the vessel's performance suddenly deteriorated due to the fouling problem mentioned earlier this must be communicated to the chartering department immediately, otherwise a commitment could be entered into which would have severe financial penalties — and we have already described the penal effects of underperformance.

A system is under development that allows a coded telex message from the vessel to be inputted directly into the office-based computer. This message would contain the 20 or so daily performance figures and could be sent at whatever intervals are deemed necessary depending on the position of the vessel from a charter fixing point of view.

If a vessel is on a long-term charter then weekly messages would suffice. If a time charter with tight performance figures is being negotiated, then daily messages would appear more appropriate. The chartering manager would then be almost in the same position as the master in knowing the performance capability of the vessel.

For those vessels fitted with satellite communications similar arrangements could be provided with enhanced round-the-clock availability, albeit at a somewhat higher cost but transmission costs are somewhat cheaper than telex.

It is an accepted fact that a satellite communication system is more satisfactory when used in conjunction with a microcomputer, normally a personal computer. This leaves the door open for a host of software programs including electronic mail, crew lists and inventory control — all of which are presently available.

12. Diagnostic and predictive programs

Planned maintenance programs have been in use for many years and need no special comment. In the natural order of things they have been updated and modified, and are generally incorporated into an inventory control and stock replenishment capability. Classification Society continuous survey of machinery computer records can now be accessed using a direct modem link into a microcomputer in the shipowner's office.

A development receiving attention is a diagnostic or predictive program which differs from the earlier planned maintenance schemes in that they are not elapsed time-based, but rely on a quite different approach.

They are probably linked to condition monitoring systems and additionally require some sort of input from an engine builder or research establishment to enable damage sensitivity criteria to be defined.

Previous problems and corrective action taken will be processed into a databank and trend analysis results from the Main Engine performance program will also be inputted. This will be mentioned in the next chapter.

13. Summary of Chapter Six

Shipboard or shore-based computers are being increasingly used to perform all manner of tasks relating to vessel operation and performance. Charterparty calculations and trend analysis techniques are ideal candidates for computer application.

In the foreseeable future most sophisticated owners will transmit data relating to vessel performance via satellite into the shipowner's office, giving instant access to current performance.

Classification Society records are now available via modem link and Lloyd's of London Press have a similar link called SEADATA giving access to all manner of particulars of interest to shipowners and charterers alike, relating to around 90,000 vessels.

The computer programs used by the author's company are very low cost and no additional monitoring equipment is needed aboard. Many expensive, sophisticated systems are available, but it is doubtful if they are cost-effective.

The programs used were developed in conjunction with UWIST (Cardiff University) who have considerable experience in maritime-related computer applications.

CHAPTER SEVEN

Remaining competitive by optimising performance

1. Introduction

The advent of the fuel crisis in 1973 was instrumental in producing very economical vessels mainly by a continual improvement in Main Engine efficiency, now probably approaching its practical limit. We must not forget that further improvements gained by ultra-sophisticated installations could be counter productive, in that higher cost shipboard personnel may be deemed necessary. The latest strategy used by shipowners in remaining competitive is to reduce manning costs by various means, most of which are outside the scope of this book. There are, however, other aspects to be considered, and these are reviewed in this chapter.

2. Dry docking

Dry dock and related costs figure high on the list of operational expenses and are probably the most important after fuel and manning, which can reverse pole position regularly dependent on fluctuations in crude oil prices.

Historically most vessels dry docked on an annual basis and classification survey rules were more or less built up around this period. In those days dry docking costs were fairly cheap and there were no undue pressures on shipowners to extend this annual dry dock. Then, around 15 years ago, dry dock capacity could barely keep up with the demand, so costs started rising and intervals between dry dockings became longer.

Although dry dock capacity is now more than adequate and prices have softened considerably, most shipowners now dry dock at two to three year intervals. Some owners go to five years which is the maximum allowed by the classification societies and to achieve this period an intermediate "in water" survey is asked for.

These intermediate surveys must be held in clear water and are performed by divers using video cameras, and the transmitted pictures are witnessed by an attendant classification surveyor.

Another measure to reduce dry dock costs is the extended interval allowed between propeller shaft survey periods. When lignum vitae bushes were used to line stern tubes, propeller shaft surveys were held at two year intervals. As white metal bushes became increasingly used and securing arrangements between propeller and shaft improved, the interval progressively increased as the reliability of the superior methods were proven.

A pilot scheme to allow a partial examination of the propeller shaft cone without fully withdrawing the propeller shaft has proven satisfactory. Considerable savings in cost can be made by deferring the full propeller shaft survey by taking advantage of this partial survey, which will then allow seven and a half years between full propeller shaft surveys.

By extending the intervals between dry docking periods it is rather important to pay close attention to the underwater paint system which has to perform a much harder task. This is particularly so with the anti-fouling paint system, and recent developments in the formulation of this product have resulted in a paint which can meet the two and a half year life nowadays considered necessary to meet the normal dry dock interval.

These paints can even meet the maximum allowable interval of five years but, of course, additional coats of paint are necessary and intermediate underwater scrubbings may also be necessary. In order to maintain the paint coatings in good condition, care must be taken in the preparation of the surfaces and high pressure water washing, grit blasting and/or sand sweeping are the usual methods employed to achieve this.

Apart from adequate paint film thickness it is important that surface roughness is kept to an absolute minimum. Frictional resistance is directly related to the condition of the underwater hull, and when the vessel leaves dry dock it should be as smooth as possible. Meters are available for measuring hull roughness, and this should be done at each dry docking in order to check that no deterioration is taking place, and also provide a basis for measuring performance.

Other items attended to at the dry docking include measurement of rudder and shaft clearances, renewal of anodes, examination/survey of sea valves and inlets and also ranging of anchors and cables. Hull steelwork damages and blemishes are also dealt with, usually with an underwriter's surveyor in attendance.

The next item highlights the difference between tanker and bulker operational scenarios. In the case of bulkers, it is normally expected that the ship's engineering staff can undertake most of the machinery survey items in service. Tanker operational patterns make this rather difficult, mainly because immobilisation of the main propulsion unit is not allowed at most tanker terminals. So the dry docking specification of a tanker will probably include many items not capable of being undertaken by ship's staff, whereas a bulker will have a much trimmed down specification.

Classification societies now allow approved chief engineers to carry out many machinery surveys on their behalf. Pilot schemes are being developed to allow certain parts of the vessel's structure to be similarly surveyed by ship's staff. Vessels altered trading patterns and reduced inport periods have made surveys rather costly, and the societies have responded to the requests of shipowning organisations by allowing these slight relaxations mentioned above. It is unlikely that items of a purely safety nature would ever be entrusted to ship's staff.

The choice of dry dock is, of course, a very important aspect in the never-ending quest of reducing operating costs closely allied to improving the commercial performance. Currency exchange rates also play an important part in the decision-making process, especially when related to the $US, the currency most frequently used for ship revenues.

Shipowners must spend a considerable amount of time drawing up a comprehensive dry dock specification encouraging ship's staff to submit detailed drawings of any parts requiring renewal.

Scheduled rates for various standard jobs should also be obtained from the various shipyards asked to quote on the specification. These would include steelwork renewals per kg, and grit blasting per M^2. When the shipyard quotes are received, the cost of any vessel deviation must be added, and if a tanker is involved the cost of the slops removed taken into consideration. Removal of slops can be an expensive exercise in certain ports and it must be taken into consideration if, of course, it is necessary to remove the slops for work to be carried out.

An idea worthy of consideration is for the shipowner to draw up his own standard conditions, and have these accepted by the shipyard. Rejection or acceptance of this type of document is allied to the prevailing market conditions, which of late favour the shipowner. Included in this document would be the definition of the word "overhaul" when used in conjunction with a piece of machinery or equipment included in the specification. It should be clearly defined that overhaul means not only dismantling and assembling but also cleaning all the parts, calibrating those subject to wear and renewal of packings and jointings.

The condition of the vessel should be the same as when delivered to the shipboard, and additional costs for clearing up what might be called the contractor's mess and debris should be for their own account.

Costs for supplying dry dock services should be defined; in the case of electrical energy it is sometimes more cost-effective to run the ship's generators than to obtain the shipyard's supply.

Legal aspects concerning guarantee, contractor's liability and payment terms should also be included especially when dealing with a shipyard for the first time. It is also a good idea to standardise the specification numbering, making it compatible with the accounts coding system, so that analysis of expenditure is made that much easier.

One unexpected result of the now accepted practice of extending dry dock intervals is the rather unknown condition of the underwater hull with respect to damaged steelwork. If major damage is found during the dry dock survey and no entry in the deck log is found that covers the incident then, after a three-year dry dock interval, it would be rather difficult to pin-point the time. This could cause problems with the underwriters, and the vexed question of the number of voyage deductibles, which would apply, is raised. Ship's staff must record any incident in the official log-book, however trivial, so that damages with unknown causes are as far as possible eliminated.

By application of the measures mentioned in this section, coupled with close supervision by experienced personnel at the dry dock, it is possible to keep dry dock costs at a much reduced level than, say five years. Part of this is due to the current depressed state of the ship repair industry which, of course, cannot be guaranteed to continue. However, shipowners have made their own contribution in containing dry docking costs to today's reasonable level.

Dry docking operations are labour-intensive, and are not suitable for the volume production techniques employed by their cousins in the shipbuilding industry. It is, therefore, difficult for them to follow the cost-cutting exercises now standard in the shipbuilding industry with the possible exception of the major ship surgery operations.

3. Voyage repairs/maintenance

Another operational expense is that incurred by voyage repairs and maintenance in the now lengthy periods between dry dockings. As vessels age, this becomes progressively difficult and, whereas ship's staff can normally overtake this function in the earlier years, they may not be able to do so in the latter part of a ship's economic life. Economic is the operative word as it may be found possible to carry out physically major renewals without too much disruption to the vessel's schedule, but the costs involved may prove uneconomical to the vessel's continued operation.

If we assume the average between dry dockings equates to around 18,000 running hours on the main propulsion unit, we can run into difficulties on those parts which cannot bear this rather long interval without attention.

This is particularly directed to tanker operations in which, as previously mentioned, immobilisation of the main propulsion unit is not normally allowed by the port authorities. A planned schedule is, therefore, called for which would highlight these critical areas and arrange for a minimum period out of service for them to be carried out. It is unusual for bulk carriers to be faced with this problem as sufficient inport time is normally available, and immobilisation restrictions are rarely met with.

To encourage ship's staff to undertake as much maintenance as possible, they must be provided with the necessary tools and material to do the intended work. In the case of older vessels facing up to wholesale pipework renewals, sufficient lengths of the correct size pipe with adequate welding materials and a pipe bending machine must be made available in order that no delays are encountered.

If the scale of the operation is beyond the capabilities of the ship's staff then recourse to a so-called flying squad should be considered. Even taking into account repatriation costs, it has been found that using a flying squad is more cost-effective than using a shipyard — particularly so in what might be called high cost areas.

It is very important that several members of the ship's engine room complement are capable of carrying out all manner of welding jobs. It goes without saying that up-to-date welding equipment with adequate lengths of cable/gas piping to reach all parts of the vessel are provided.

A good selection of workshop machinery is essential if ship's staff are to help reduce dependence on expensive shore-based establishments. Training should be given in the overhaul of fuel pumps and injection valves and specialised lapping equipment obtained to facilitate this delicate type of work.

There are many other items of auxiliary equipment which must be kept in good order to maximise performance. Some of these tend to be overlooked, even though pressure and temperature sensors are generally fitted to indicate if they are functioning correctly.

Included in this category are the filters provided in the various main and sub-systems to protect delicate parts from the ingress of harmful particles. Failure to keep filters in good condition can lead to all sorts of operational problems, and the importance of maintaining their operational efficiency cannot be emphasised too strongly.

Other important pieces of equipment are the various heat exchangers which are used for such diverse purposes as heating the fuel for the Main Engine, boiler, generators — also for preheating the fuel prior to purification. They are also used for cooling scavenge air, jacket water and lubricating oil — also for condensing steam.

Any system malfunction resulting in high temperature conditions will probably cause precipitation of harmful impurities onto the heat exchanger surfaces resulting in loss of efficiency. If the system temperature is not quickly restored to acceptable limits, serious damage can result.

In the case of main scavenge air coolers, an inbuilt chemical cleaning system is standard and should be used as recommended by the engine builders. The other heat exchangers should be cleaned when system operating conditions indicate fouling is starting to take place.

Turbo-chargers are another critical item in optimising performance and must be kept in good condition. They are subject to fouling, both on the gas and air sides. Arrangements are provided to water wash the gas side, and the air side can be similarly dealt with using a mild detergent. It is also possible to use a dry-cleaning technique on the gas side using rice or ground-nut shells instead of water. Dirty turbo-chargers have a knock-on effect on performance in that inefficient turbo-chargers provide insufficient combustion air to the engine, which results in imperfect combustion hence unburnt particles depositing themselves on the turbo-charger blades. If the situation is not restored quickly, a quite serious state of affairs can develop leading to all manner of problems in the combustion process.

Most of the subject matter that we have considered in this section relates to good housekeeping and this, of course, is a prerequisite of maximising performance.

There are other parts of the machinery to consider; for example, diesel generators, boilers and the electrical supply and distribution system.

Diesel generators require similar maintenance to that exercised on Main Engines and require no special comment except that in a well-designed engine room a diesel generator will always be available for overhaul at sea. This allows the ship's engineering staff sufficient time to undertake preventative maintenance tasks at sea, thus ensuring that the generators are in tiptop condition for cargo duties in port.

Boilers need to be kept clean both on the water and combustion sides, and are unlikely to survive between the two-yearly survey period without attention. Careful control should be kept on the boiler water and, in addition to the daily on board tests, the suppliers of the boiler water treatment should be asked for regular advice on the test results. Boiler corrosion can develop rapidly if the water is not kept in good condition and any signs of this not being the case should be followed up immediately.

One of the casualties in staff cutting exercises has been the ship's electrician and, except for vessels having complicated electrical systems, the electrician is nowadays rarely seen. It is a good idea to train the senior engineers in the rudiments of practical electrical knowledge so that they can deal with emergencies developing in the electrical systems.

Since a.c. replaced d.c. current 20 or so years ago, the preventative maintenance of electrical maintenance has decreased — especially in the case of generators and motors. Even so, routine electrical maintenance duties should be carried out by the engineers trained in this task.

4. Condition monitoring

Of the machinery so far discussed in this chapter, we have not yet mentioned pumps, fans and other rotating equipment. One means of reducing the maintenance work-load on this type of equipment is to introduce condition monitoring in the form of vibration analysers and shock pulse meters.

Readings taken on a regular basis can be plotted on a time basis and indications of impending problems can be averted by taking the appropriate action. Classification societies will accept this type of condition monitoring as an alternative to opening up the item for survey. Condition monitoring is only in its infancy when applied to the marine engineering industry, unlike many shore-based industries which use it extensively.

Sophisticated condition monitoring systems can be used to measure wear on cylinder liners whilst the engine is running, they can also measure pressures and temperatures in the combustion process as a means of defining combustibility to name but a few applications of their use. To be accepted they have to be cost-effective, and this aspect is presently being investigated by various professional institutions.

5. Planned maintenance

Planned maintenance systems became popular around 15 years ago and probably owed their popularity to the lack of continuity of serving engineer officers caused by unavailability of personnel. This made the previous method of committing the performance of each item of machinery to the then captive engineer's memory impossible. These systems can become paperwork generators inundating the shipowner's office with hundreds of forms he could well do without. When used in conjunction with a stock control and replacement ordering scheme these systems are very useful, providing the shipowner has the necessary checking procedure to intercept excessive overstocking.

It has been found beneficial to replace calendar-based intervals by running hour intervals, and cheap running hour meters have been connected to those pieces of equipment previously having unknown running periods.

Another beneficial move has been to allow those items of equipment due an overhaul or service under the planned maintenance system to be checked under working conditions, and if working satisfactorily the period can be extended. Items not normally capable of being unduly extended would include Main Engine pistons and turbo-chargers, and a fully developed condition monitoring system could help in these or similar cases.

Operational efficiency is dependent on the optimised performance of the vessel and its machinery, these two prerequisites could easily be stated in reverse order and still have the same meaning; the point being that whatever means are applied they must always be technically and commercially acceptable.

6. Unplanned maintenance

This particular aspect of the vessel's operational scenario is rather difficult to predict. Even in a well run efficient vessel, events can occur which will interfere with the schedule, and also involve considerable cost. Most of these incidents will involve claims on the owner's hull and machinery policy and, such is the scope of the range of damages under this heading, it is difficult to single out any typical example.

One particular aspect which can progressively develop is the deterioration of steelwork due to the normal corrosion process which, of course, is not covered by standard hull policy clauses. This can be visible in the case of the upper deck or out of sight as in the case of ballast tanks or cargo spaces.

The supply of a grit blaster and an adequate supply of grit can produce remarkable results with a well-motivated crew in a comparatively short time, given favourable weather conditions. This operation is ideally suited to the upper deck and hatch coaming area, but certain restrictions on their use in tanks and holds will, of course, apply.

Corrosion in ballast tanks, once started, can develop rather quickly and an adequate supply of sacrificial anodes can prevent costly steelwork repairs if applied at the onset of corrosion.

Cargo gear, especially on tankers, is another item which can create havoc with the vessel's schedule. Cargo pumps, in many instances, rarely receive the priority treatment they deserve. Nowadays diesel engineers account for a majority of certificated ranks and the steam certificated engineer is fast disappearing. The steam driven cargo pumps with horsepowers approaching that of some earlier Main Engines require careful attention and diesel engineers must be trained to give this otherwise expensive incidents will arise.

This could also apply to deck cranes which nowadays are generally electro-hydraulic and require specialised maintenance techniques not normally acquired by seagoing engineers in the normal course of their career. Training courses at the manufacturer's works are strongly recommended if expensive breakdowns are to be avoided.

7. Spares and supplies

Expenditure under this heading accounts for a considerable proportion of direct operating expenses with spare part replacement probably taking the lion's share especially if the cost of air freighting these parts is included.

A critical examination of the actual stocks of spares held aboard some time ago did reveal that on average more parts than those disclosed in the inventory returns were held. It should be mentioned that the number of different items included in a complete inventory is quite large, the physical check time consuming and regular repeat examinations are therefore unlikely to be held.

This is an area in which strict controls can result in quite significant savings being made. One of the major faults is the more or less automatic ordering of replacements when parts are taken from the inventory into use.

When freight rates are high, and off hire periods can prove to be costly, it is rather important that spare part inventories are kept at a high level. When rates are low there is not the same importance in keeping to this philosophy, and some degree of risk can obviously be taken.

Most items of auxiliary machinery have a stand-by unit automatically brought into use when the on-line unit fails. This, in fact, is a Classification Society requirement for unmanned machinery space operation.

Given that the spare parts inventory will probably include the moving parts of the particular piece of equipment we have in mind, it follows that no less than three sets of parts subject to wear are carried for most auxiliary units. Obviously some sort of compromise can be made taking into account that most shipowners are fighting for survival.

With respect to deck, engine and catering supplies — which include food — some sort of control should be exercised, whether it be by giving the master of the vessel a monetary spending limit or by screening the indents at head office.

When sterling is weak, it has been found to be cost-effective to ship supplies to vessels on the European Continent from the UK and, in fact, this practice is extensively used.

A final point concerns airfreighting heavy machinery parts which, of course, is a very expensive operation. Ship's staff should be made aware of this high cost factor and should be encouraged to plan overhauls, possibly requiring parts renewal, at ports having full spare part facilities; for example, Rotterdam and Yokohama. It is realised that this is not always possible, but efforts made in this direction can produce significant savings.

8. Lubricating oil

Lubricating oil is a costly commodity and figures high on the list of operational expenses. A technical approach can be made to contain these expenses at what might be called acceptable levels of consumption.

Starting with cylinder oil there is a distinct relationship between the amount of cylinder oil used and rate of wear on the cylinder liner walls. By supplying copious amounts of cylinder oil the rate of liner wear is much reduced, but a commercial viewpoint is necessary in order to consider all the facts.

Cylinder oil consumption is usually measured in grams per horsepower hour (g/b.h.p./hour) and liner wear rate is measured in millimetres per 1,000 running hours (mm/1,000 hours).

For an example we could say that a cylinder oil feed rate of 0.7 g/b.h.p./hour gives a liner wear rate of 0.1 mm per 1,000 hours. On a 700 mm bore engine the maximum allowable wear is usually judged to be 5.5 mm so that the life of the cylinder liners would be:

$$\frac{5.5}{0.1} \times 1,000 = 55,000 \text{ hours}$$

say 9.2 years.

If we increased the feed rate to 1.0 g/b.h.p./hour we could, perhaps, reduce the liner wear rate to 0.08 mm/1,000 hours giving a liner life of:

$$\frac{5.5}{0.08} \times 1,000 = 68,750 \text{ hours}$$

or 11.4 years.

The cost of the increased oil can now be calculated. We will assume the engine in the example is 10,000 b.h.p. running 6,000 hours per year. With a feed rate of 0.7 g/b.h.p./hour the annual amount (kg) of cylinder oil consumed will be:

$$\frac{0.7 \times 10,000 \times 6,000}{1,000} = 42,000$$

and with cylinder oil at $0.70 per litre having a density of 0.96, the annual cost would be:

$$\frac{42,000}{0.96} \times \$0.70 = \$30,625$$

The annual cost of the increased feed rate would be:

$$\frac{1 \times 10,000 \times 6,000}{1,000 \times 0.96} \times \$0.70 = \$43,750 \text{ hours}$$

The question to be answered is can we justify paying $13,125 per year in additional cylinder oil to extend the life of the cylinder liners 2.2 years? The answer is probably "no", but it could depend on the shipowner's intentions about whether to sell the vessel and, if so, when.

What can be learned from this simple exercise is that an increase in the cylinder oil feed rate of 0.1 g/b.h.p./hour costs:

$$\frac{\$13,125}{3} = \$4,375 \text{ per year}$$

and must be balanced against any improvement in cylinder life, itself dependent on the owner's sales policy.

In the case of crankcase oil, the consumption is not a feed rate, as is the case for cylinder oil, but is really a loss rate caused by a variety of reasons. These include leakages from crankcase doors and piston rod glands, evaporation losses and losses in the purification system.

A well looked after machinery installation will have a crankcase oil consumption of around 0.3 g/b.h.p./hour. Because crankcase oil is cheaper than cylinder oil an increase of 0.1 g/b.h.p./hour in the example given earlier would have an annual cost of around $3,325 per year.

A brief word on the performance of these commodities may now be appropriate. Cylinder oil has a joint function in that it lubricates the cylinder liner piston ring interface and additionally neutralises the acidic products of combustion in this hostile area, thus preventing corrosion.

As Main Engine pressures and temperatures are increased to optimise thermal efficiency, the function of cylinder oil is approaching its limit in a hydrocarbon form. Research is proceeding to produce a pure synthetic or synthetic-based cylinder oil capable of meeting the rigorous requirements now demanded in order to contain cylinder liner wear at an acceptable level — which is proving difficult on these new highly rated long stroke engines.

The consumption of lubricating oil in diesel generator engines must also be considered, even though it is usually less than the lubrication consumption of the Main Engine cylinder and crankcase oil. Because of the comparatively low powers developed in diesel generators, it is not usual to relate the lubricating oil consumption to power as in the case of Main Engines, but simply to state it in litres per day. On a modern diesel generator engine we can expect a lubricating oil consumption in the range of 20 litres per day.

So how can we monitor the performance of lubricating oil? A simple table as indicated is considered sufficient for most purposes:

Table 14.   Lubricating oil performance

Vessel	Average output ME %	Liner wear		Cylinder oil consumption		Crankcase oil consumption		Diesel generator consumption	
		84/5	85/6	84/5	85/6	84/5	85/6	84/5	85/6
A	65	0.07	0.07	0.7	0.6	0.3	0.3	25	24
B	71	0.08	0.07	0.6	0.6	0.4	0.4	21	19

Vessels can be grouped into the various oil suppliers and average results compared with each other if more than one supplier is used. Target values for the liner wear and tabled consumptions can also be shown.

It is also possible to monitor the rate of cylinder liner wear indicating maximum liner wear readings on an elapsed time basis as shown in Figure 32. This gives an accurate indication of cylinder liner wear and the slope of the individual cylinder curves will indicate changes in the wear rate pattern. Feed rate changes, or even changes in the supplier of the oil, can be introduced and the result of these changes shown at subsequent cylinder calibrations.

Using this approach it was possible to demonstrate the rather remarkable increase in the rate of liner wear occasioned by the use of fuel with a high level of naphthenic acid and helped convince underwriters' surveyors of the creditability of our claim.

The condition of the system lubricating oils used in Main Engines and diesel generators are subject to vigorous attacks, usually connected with the combustion process. Lubricating oils must have an alkalinity reserve to combat the ingress of acidic particles, and regular checks must be made on the oil's condition. Simple test kits are provided by the oil companies for onboard testing and, in addition, regular more thorough tests are made in the oil companies' laboratories. These laboratory tests include sophisticated metallic particle counts which will indicate if any excessive wear is taking place on the metallic bearing surfaces and ideally will avoid serious breakdowns.

We have seen how by monitoring the consumption of the various lubricating oils we can effect savings, we can also monitor the performance of different suppliers' oils, but this is not normally immediately apparent and some time must elapse before any oil's superior performance is established. This would probably involve at least one piston

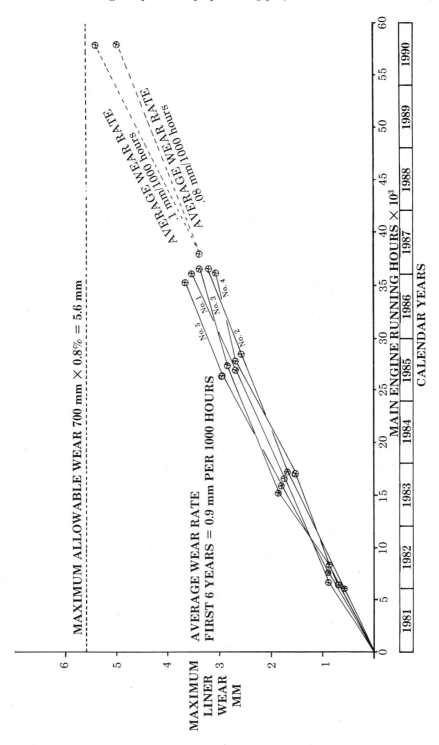

Fig. 32. Cylinder liner wear graph

overhaul during which cylinder calibrations are taken and a two-year period could be involved.

9. Insurance and P & I

Premiums and calls paid by shipowners under the general heading of insurance are also a considerable part of direct operating expenses. They are directly related to what might be called the performance of the shipowner as reflected by the extent of his claims on underwriters and P & I clubs. Obviously the more efficiently an owner operates his vessels so the incidence of claims will be reduced, and this will be reflected in his premiums for the following year.

A significant amount of major hull damage claims concern groundings, and it is very important that up-to-date hydrographical charts are supplied to the vessels and that these are kept up-to-date by making corrections as regularly advised by the authorities.

In a recent arbitration it was ruled that shipowners have the ultimate responsibility in ensuring that this aspect is fully complied with, and negligent acts by ship's staff do not abrogate this responsibility. The significance of this ruling is far-reaching in that insurance cover for owner's negligent action must be covered — probably by the owner's P & I club.

If we assume that most hull damage claims can be attributed to a known incident, owners do not have any problem stating a cause. As previously mentioned, this does not include those damages discovered in dry dock which may be delegated into a date and cause unknown category if log entries are not sufficiently detailed. Shipowners are then left with the problem of presenting underwriters with a cause when machinery damage claims are being examined by underwriter surveyors.

The main problem is to identify the cause with a relevant clause in the hull and machinery policy. These clauses do vary dependent on where the insurance cover is placed, whether it be London, New York or Gothenburg to name but a few of the more often used insurance centres.

It is in the shipowner's interest that the clauses applicable to his fleet are known to those members of his staff dealing with claims, and this should include the master and chief engineer.

Machinery damages are in some instances simply due to fair wear and tear and are, therefore, not covered by any standard hull and machinery policy clauses. It is, however, very important that each incident is fully investigated by competent personnel so that all avenues are exhausted before it is placed in this category.

Speedy settlement of legitimate claims is in the shipowner's interest as he has already paid the contractor for carrying out the work, and lengthy delays in dealing with the adjustment of the claim can be costly.

The question of seaworthiness is sometimes brought up, especially when cargo underwriters' interests are concerned, for example, when a vessel is diverted to a port of refuge due to damage and general average is declared.

Normally the costs involved in general average, which can include substantial towage costs, are shared by the hull and machinery underwriters and cargo underwriters on a pro rata respective value basis.

If cargo underwriters can show that the vessel involved was at the commencement of the voyage in question unseaworthy then they will probably refuse to contribute to their portion of general average expenses. It is very important, therefore, that shipowners ensure that in all respects their vessels are seaworthy at the commencement of a voyage if they are not to be placed in this position.

Finally, if vessels are out of service or subject to excessive port delays this information should be communicated to the underwriters for a possible reduction in premium payment.

10. Miscellaneous charges

The final category of operational expenses comes under the heading of miscellaneous. These will normally include radio traffic, agents' disbursements, various fees and, perhaps, a management fee in the case of managed vessels.

Radio traffic costs are usually monitored by one of several specialised firms who, for a nominal sum, will look after the commercial interests of the shipowner. In recent years telex and satellite communication systems have improved the performance of message transmission between the vessel and the owner/operator. On time-chartered vessels those messages sent which are related to charterers' activities should, of course, be paid by them.

Agents' disbursements require careful vetting by experienced personnel in order to properly apportion the various charges. Agents generally ask for, and usually receive, funds to cover owners' expected outlays during the vessel's visit. This also is an area requiring careful control.

Such is the nature and scope of the subject matter included in this section that it is very difficult to generalise. It does appear that, in some instances, the control of miscellaneous charges is not given the attention it deserves.

11. Summary of Chapter Seven

We have learned how dry dock costs have effectively been reduced mainly by lengthening the interval between dry dockings. Various other measures open to shipowners to make them more competitive are described.

CHAPTER EIGHT

Past, present and future

1. Introduction

Most of the various factors affecting performance of existing and proposed new buildings have been reviewed. We should now stand back and look at the performance of past vessels and review this against current designs and possibilities for the future.

2. The past

As we have generally confined our comments to what might nowadays be called conventional vessels, we will only go back in time to when the situation started to change and economies of scale played a big part in the development of new designs.

During World War II, and immediately after, the then standard vessel was a 10,000 TDW tween deck dry cargo bulker as typified by the large number of wartime Liberty, Fort and Empire vessels built during this period. Tankers tended to be slightly larger, probably on account of the fact that they did not use enclosed dock systems with their severe restrictions on vessels' dimensions.

The machinery on these wartime-built vessels was invariably steam reciprocating with oil fired boilers, the Liberty ships had an added attraction in being supplied with water tube boilers instead of the Scotch boilers used extensively during the period.

Performance was not a consideration when ordering a vessel in those days. They were all of a muchness having a loaded speed of around 10 knots and a consumption of about 35 tons fuel oil. Coal burners were still to be occasionally found and their consumption rose to around 50 tons per day for a similar performance.

A small percentage of vessels were fitted with diesel engines and their consumption would be around 12 tons of diesel oil. Heavy fuel oil was not to become popular until the 1950s and diesel oil was used more or less exclusively during this period.

Towards the end of the war the Victory design came into service, these vessels were turbine driven and had somewhat larger deadweights and higher speeds. The T2 tanker also came into service during this period and represented a major step foward in that they were provided with turbo-electric machinery with the main propulsion unit driving the propeller at sea, and the cargo pumps whilst in port.

Peacetime designs for bulk carriers followed the wartime standard vessels in that they were still tween deckers, but an increasing number of diesel engines were being specified.

British and Norwegian fleets then headed the league table of shipowning nations and Doxford, Harland & Wolff (B&W) Sulzer and Gotarverken engines gradually replaced steam engines.

In the case of tankers, steam turbines were also popular — especially with British and USA fleets, both of whom were rather dominant in the oil industry. Norwegian owners tended to favour diesel propulsion and kept to this choice even when deadweight capacity started to rise in the mid 1950s.

Self-trimming bulk carriers also became popular around this time and the time-consuming practice of erecting shifting boards on tween deckers when carrying grain cargoes became a thing of the past.

As tankers and bulk carriers started their ever-increasing size leap-frogging tactics steam turbines occasionally were favoured if only because diesel engine designs could not keep pace with the increased outputs demanded by the higher deadweights. Even as late as 1973 steam turbine VLCCs were keeping up with diesel designs but the oil crisis in the autumn of that year was eventually to change all that.

This was the first of several oil crises which saw bunker prices rise from $20 a tonne to $80 then $180 at which point it stabilised for around five years or so. These price hikes sounded the death knell for the steam turbine and even though oil prices have now softened, it would appear unlikely that steam turbines will reappear in any large numbers.

When oil prices are high, ship's speed is a very expensive commodity and horsepowers are now, therefore, much lower — which is another point against the reintroduction of the steam turbine.

There was recently some interest shown in coal-burning steam turbine vessels and several were built in Italy and Japan. They owed their short-lived popularity to a period when oil prices peaked at nearly $200 a tonne and coal prices became competitive if the additional capital costs for these rather special ships were largely ignored.

The oil price hikes, as well as dealing a death blow to the steam turbine, also focused attention on the thermal efficiency of current diesel engine designs — then running at around 40%.

One of the problems with the then current diesel engines was their rather high revolutions. The first diesel VLCCs had propeller revolutions of 100-110 RPM whereas the steam turbine VLCCs' propellers turned at around 70 RPM which made them more efficient from a propulsive point of view.

Several slow-speed diesel vessels using gearboxes to reduce the propeller revolutions were designed but, as far as is known, none were built. It should be mentioned that medium-speed diesels are invariably geared down and have always had this advantage over their slow-speed rivals.

The derated slow-speed long stroke diesel then made its appearance and immediately the high revolution problem was solvable, albeit at additional cost. Therefore, a diesel VLCC ordered with such an engine could have a propeller turning at even less revolutions than the 70 RPM of the steam turbine.

Other technical developments during the period under review mainly revolved around energy conservation measures occasioned by the high fuel costs.

The size of tankers stabilised at around 540,000 TDW in the late 1970s even though designs for 1,000,000 TDW tankers were produced and several shipyards have the capabilities of building these mammoths.

Commercial reasons, rather than conservation measures, were largely the cause of the decline in the tanker market and it is only very recently that replacement VLCCs have been contracted for. It would appear unlikely that VLCCs will ever again become popular, purely for commercial and *not* technical reasons.

Bulk carriers are still increasing in size and the current maximum deadweight is around 350,000 TDW. These vessels are generally designed for specific trades; for example, iron ore from South America to Japan.

Returning to diesel engine design, the thermal efficiency had — as previously mentioned — stabilised at around 40% ever since the introduction of two stroke turbo-charging in the early 1950s. Thermal efficiencies have steadily improved, mainly by increasing the degree of turbo-charging which leads to higher cyclic pressures and temperatures and, of course, efficiency — and efficiencies are currently around 53%.

Automation of machinery spaces started gaining popularity in the early 1960s and eventually led to the unmanned engine-room. In the beginning no serious attempt to substantially reduce manning as a result of introducing automation was made, probably because manning costs were not a problem. The recent prolonged trough the market is experiencing, however, has directed attention to manning costs. Some of the high crew-cost shipowning nations, notably Scandinavian ones, have made significant reductions in manning numbers enabling nationals to be employed in the surviving berths. Even this drastic action does not always appear to be sufficient. Flagging out and offshore agreements now have the same sort of press coverage energy conservation had a decade ago.

3. The present

This brings us up to the present which, from a commercial point of view, is decidedly poor. But there appears to be a slight improvement (fourth quarter, 1986).

Not so many new buildings are being contracted for and most shipbuilding nations are reducing capacity in an attempt to weather the storm. The overlying trend in world-wide seaborne trade is not declining significantly, but too many vessels were built on purely speculative grounds and, until equilibrium is restored, the market is unlikely to substantially recover.

Present-day designs are now so efficient that fewer vessels are required, and this could be a factor in the scrapping rate of older vessels; for example, as already seen with steam turbine VLCCs.

Most surviving shipowners have made strenuous efforts to remain competitive by building efficient vessels and reducing operating costs on the lines mentioned in previous chapters.

European flag vessels have been reduced in numbers since the days when the UK and Norway headed the league and these former shipowning powers are placed fifth and ninth respectively, with every possibility of them falling further.

Shipbuilding costs in mid-1986 were the lowest for many years, although there was a period in 1978 when they were lower. A VLCC cost around $16m in 1968. This gradually rose to $40m in 1974. In the early 1980s it had risen to around $70m and in the third quarter, 1986, the price is around $40m — but with every chance of it rising if the recent flurry of ordering activity is continued.

One unfortunate fact is that any revival in the freight market can only last as long as the now hungry and under-utilised shipyards can move back into top gear.

Many European officers and crew have been replaced, usually by lower cost Far Eastern employees. The haste with which this transfer is being accomplished is quite remarkable and the situation of shortages in key ranks which was prevelant in Europe in the last decade could conceivably reappear.

A typical present-day vessel will, of course, have the latest fuel efficient Main Engine and most of the more popular energy conservation features. Registry will probably be Monrovia, Hong Kong or other currently fashionable ports with the owner's name not immediately recognisable as coming from a familiar shipowning stable. Officers and crew will probably be a mixture of nationalities and supplied by an agency having its office in the Far East. The nationality of the master may give a clue to the identity and location of the beneficial owner, but this cannot always be relied on.

The performance of the vessel should be at least 30% better than that of a similar vessel built 10 years ago. Advances in technology will mean that less maintenance is required and fewer staff needed to operate her. This coupled with low shipbuilding costs and perhaps favourable finance terms make it very attractive for speculators to move in and order newbuildings even though the initial return on investment looks bleak.

4. The future

Viewed from the midst of the longest recollectable recession in the shipping industry, it is rather difficult to predict the future without becoming influenced by the current gloom. It can only be said that those shipowners who do survive will be leaner and fitter, to quote a well-known phrase.

Vessels delivered in the last three years or so will almost certainly be to the latest fuel efficient design having various features for reduced manning. With respect to fuel efficiency, it is rather difficult to foresee what significant improvements can be made to the already extremely efficient plants currently available.

It is possible that with the use of improved materials cycle combustion temperatures and pressures can be further increased, leading to a maximum improvement in thermal efficiency of around 5%.

On suitable vessels, additional power turbo-chargers can be incorporated with an effective contribution in thermal efficiency of, perhaps, 4%. It would be imprudent to add these two possibilities together, as the improvement in cyclic efficiency may lessen the possibility of incorporating additional power turbo-chargers. This is what happened to the sophisticated exhaust gas heat recovery scheme which became unsuitable, except for high powered vessels, as a result of improved thermal efficiency of the main propulsion unit.

Propeller design has not drastically altered for many decades except, of course, for the recent improvements mentioned in Chapter One. Indeed, there does not appear to be any real development work being undertaken on any revolutionary new ideas apart from an electromagnetic thrust system employing similar principles to the linear motor.

The basic problem in hydrodynamics is the penal effect of the cube law relationship, which effectively prohibits high-speed operations because of the energy costs. Under this relationship, a 20% increase in ship's speed requires a 60% increase in fuel consumption. The conventional screw propeller and its derivatives are rather efficient at low ship speeds so, whilst fossil fuel is being used, it would appear unlikely that any major conceptual changes in propulsion systems will be developed. It would therefore appear that until fossil fuel is replaced by something much cheaper, we will be left with fairly conventional propulsion systems.

To decide at what point in time fossil fuel, in the form of oil, will in fact run out is what might be called the $64,000 question. As recent as 1980 we were advised by well-informed sources that this would be in about 20 years.

Because of the huge switch from oil to coal by the really big consumers, i.e. power stations, also the effect of energy conservation measures and the gradual increase in alternative power sources — for example, nuclear and hydro — this gloomy forecast has happily proven to be incorrect.

Another major switch not yet realised but equally effective will be the gradual use of diesel engines in automobiles. Probably the biggest single product from the crude barrel is used in automobiles and the effect of a virtual 10% decrease in consumption by using more efficient diesel engines must be considered.

It is comparatively easy for major utilities like a 3,000 megawatt steam turbine power-station, which incidentally is equivalent to four million horsepower, to switch from oil to coal, thus saving around seven million tonnes of oil per year. It is a rather different story to adopt this switch on a moderately powered vessel of say 10,000 b.h.p. which, unlike a power station, will have a diesel engine — not a steam turbine.

Whereas land-based power-stations can switch from oil to coal and then back to oil if price differentials demand this course of action, it is not possible to adopt this approach unless a steam turbine vessel was involved.

To make a major decision involving a steam turbine instead of a diesel engine would require a long-term accurate forecast of the cost differential between oil and coal.

The availability of world-wide supplies of coal with suitable properties would present insurmountable difficulties, although vessels on dedicated coal trades would not have this problem. Those coal-burning vessels mentioned earlier in the chapter were built with this in mind.

We could say that, subject to the long-term stability of coal prices being assured, there is a possibility of coal fired dry cargo vessels having a future in the medium-term if and when oil shortages reappear.

Providing the same level of installed horsepower is still acceptable there is also a possibility of reciprocating steam engines replacing the steam turbines now currently specified.

These modern reciprocating steam engines are of a somewhat lower thermal efficiency than the steam turbine, but offer reduced operational difficulties and require a lower level of expertise to operate.

It could be that the standard dry cargo vessel of the future is propelled by a coal fired steam reciprocating engine having a rather poor thermal efficiency but an efficient commercial performance in terms of cents per horsepower produced.

At this moment in time (fourth quarter, 1986) a modern diesel engine using $70 per tonne heavy fuel oil at 53% thermal efficiency will have a commercial performance of 0.83 cents per horsepower hour.

Should heavy fuel oil costs return to $180 per tonne this will rise to 2.12 cents per horsepower hour. Consider now, a steam turbine using heavy fuel oil at $70 per tonne having a thermal efficiency of around 32%, the commercial performance would be 1.4 cents per horsepower hour rising to 3.6 cents per horsepower hour with fuel at $180 per tonne.

If we then consider the coal-fired steam turbine alternative, we still have a 32% thermal efficiency, but we must now take into account the reduced heat value of coal.

With coal having a calorific value of only 25 megajoule/kg against 40 megajoule/kg for oil, we obviously must burn more coal than fuel oil in the approximate relationship of 40/25 or 1.6 times more coal than oil.

We must also take into account the lower cost of coal which, unfortunately, is not so easy to find as is the cost of fuel oil and seems to have widely fluctuating costs on a world-wide basis. For the sake of reference we will use a coal cost of $35 per tonne which appears to be realistic for the fourth quarter, 1986.

The commercial performance of the coal fired turbine using these figures would be 1.12 cents/hp/hour. It is rather difficult to find the thermal efficiency of a modern steam reciprocating engine, but it is understood to be around 28%. If this is correct then the commercial performance of a coal-fired reciprocating engine would be around 1.30 cents per horsepower hour.

All this can be shown in a simple table as illustrated in Table 15.

Table 15.   **Performance costs — various propulsion systems** *(commercial performance in cents/b.h.p./hour)*

Main propulsion system	Oil-cost $70 tonne	Oil cost $180 tonne	Coal cost $35 tonne	Coal cost $70 tonne
Modern diesel engine	0.83	2.12	—	—
Oil fired steam turbine	1.40	3.60	—	—
Coal fired steam turbine	—	—	1.12	2.24
Oil fired steam reciprocating engine	1.60	4.10	—	—
Coal fired steam reciprocating engine	—	—	1.30	2.60

In studying this table we must consider that it would appear unlikely that steam cycle efficiency could be improved, whereas we have the possibility of increasing diesel cycle efficiency by 5% — 8% in the medium-term.

We must also consider the new building capital cost element, which currently favours diesel propulsion, but in a crisis such as that occasioned by a serious interruption of oil supplies could be ignored.

There are also alternative fuels to consider which may prove to be more attractive than coal with its rather difficult handling arrangements. The first that springs to mind is the direct conversion of the abundant reserves of coal into oil by various means already in use. This brings in the possible use of huge amounts of brown coal for this purpose which is not particularly suited for either thermal or metallurgical purposes. The only drawback in converting coal to oil is the rather high conversion costs which will no doubt be reduced as techniques improve and volume production increases.

If oil costs rise to upwards of $300 per tonne it would appear that direct conversion of coal to oil or the use of coal itself could be an attractive alternative to oil.

Another possibility is the use of methanol, methane and ethanol, all of which can be produced in sufficiently large quantities by various processes and could possibly be used by suitably adapted diesel engines. This would avoid the fuel handling problems occasioned by the use of solid fuel in the form of coal.

Recent propulsion systems have incorporated the use of wind power, usually as a contribution rather than as the sole source of power. In certain applications they do make a positive contribution to the overall economy, but whether the capital cost coupled with the additional labour required for sail operation can be justified is the subject of much debate. It would appear unlikely that these would catch on from an international basis in that low cost Far Eastern operators would not be favourably impressed.

This more or less covers the alternate power sources with the obvious omission of nuclear energy. Many hundreds of nuclear propelled vessels are in service in the form of naval vessels where commercial considerations take a poor second place to those of endurance and redundancy not normally a consideration in the merchant marine.

Several nuclear merchant vessels have been in service and we exclude Eastern Block ice-breakers from this observation. The obvious detractions from the acceptance of this form of energy is the immense capital cost, the necessary expertise and numbers of operating personnel required — and the sometimes hysterical objection to its use on environmental grounds. It is exceedingly unlikely that this form of propulsion will be used in the merchant marine service in the short- to medium-term. There is a distinct possibility that it will be in universal use in the long-term when fossil fuel and its various derivatives are commercially unattractive.

One advantage in the use of nuclear propulsion is that the previous dependence on observing the energy related restriction of the cube law would largely disappear. Having met the high capital cost of a nuclear propelled vessel, the unit fuel costs are not really significant and this givs the opportunity of rather high powers and subsequent higher ship speeds than those currently practised.

5. Summary of Chapter Eight

We have seen how diesel propulsion gradually took over from steam, leading to the virtual extinction of steam propulsion in the early 1980s.

The thermal efficiency of the diesel engine has steadily risen from 40% to the current 53%, and has a possibility of reaching a maximum of, perhaps, 58% in the medium-term.

Alternative fuels in the form of coal or oil produced from coal appear likely candidates in the medium-term, but nuclear propulsion could be the sole source in the long term.

Conclusion

The tremendous improvement in the performance of ships over the last decade has made the cost of sea transport much cheaper. Even so, most shipowners of the type of vessel reviewed in this book are finding it difficult to stay in business.

A 10-year-old diesel VLCC travels only 5.4 miles per tonne of fuel consumed, a modern VLCC will travel 7.7 miles per tonne. This 40% improvement in performance is typical for all the ship types reviewed in this book. From a technical point of view there is still the possibility of improving the performance of a modern vessel and means of doing this have been outlined. However, in the short term there is not the same amount of scope for further improvements of the magnitude enjoyed over the last decade. Therefore, most owners have directed their attention to the reduction of direct operating costs and currently manning is under scrutiny.

Flagging out, offshore agreements and outside management contracts are now the vogue and manning costs have been reduced substantially by employing various combinations of these. Some shipowners have retained their National Flag with personnel greatly reduced in number but highly trained and Panamax size vessels can be manned with a total complement of 18 using this approach. This would compare with around 28 when using a Far Eastern low cost management company.

Another possibility for reducing operating costs gaining in popularity is to write down the ship's value to reflect a more accurate assessment of its current value thus reducing depreciation costs. This specialised subject is, however, outwith the scope of this book. In a similar manner the insured value can also be reduced thus effecting a saving in insurance premium payments. It is also possible to increase the deductible value of each casualty accident which can result in savings in a well run fleet but could be counter productive in a not so well run fleet. By employing these various economies direct operating costs can be reduced from around $4,000 to approximately $2,500 per day and even lower figures can be obtained at the expense of having a less efficient operation.

The large amount of surplus tonnage is the main reason that freight rates are depressed and no positive improvement can take place until this surplus is absorbed in some way or other. Owners of modern efficient ships with low operating costs will obviously stand a better chance of capitalising on this when the market recovers.

Up until the present time (fourth quarter, 1986) the freight market shows no signs of recovery with the possible exception of the tanker section. It has previously always staged a recovery for reasons not apparently obvious at the time. If and when the recovery comes, those owners who have taken steps to optimise performance should have a reasonable future.

Index